Plasticity, Robustness, Development and Evolution

How do we understand and explain the apparent dichotomy between plasticity and robustness in the context of development? Can we identify these complex processes without resorting to 'either/or' solutions?

Written by two leaders in the field, this is the first book to fully unravel the complexity of the subject, explaining that the epigenetic processes generating plasticity and robustness are, in fact, deeply intertwined. It identifies the different mechanisms that generate robustness and the various forms of plasticity, before considering the functional significance of the integrated mechanisms and how the component processes might have evolved. Finally, it highlights the ways in which epigenetic mechanisms could be instrumental in driving evolutionary change.

Essential reading for biologists and psychologists interested in epigenetics and evolution, this book is also a valuable resource for biological anthropologists, sociobiologists, child psychologists and paediatricians.

PROFESSOR SIR PATRICK BATESON FRS is Emeritus Professor of Ethology at the University of Cambridge. He is President of the Zoological Society of London, and former Biological Secretary and Vice-President of the Royal Society. He has a long-standing interest in behavioural development and in evolutionary theory.

PROFESSOR SIR PETER GLUCKMAN FRS is arguably New Zealand's most recognised biomedical scientist. A University of Auckland Distinguished Professor, he is Head of the Centre for Human Evolution, Adaptation and Diseases and former Director of the Liggins Institute for Medical Research. The bulk of his recent research has related to developmental plasticity and its relationship to human health.

Plasticity, Robustness, Development and Evolution

PATRICK
BATESON FRS
University of Cambridge, UK

PETER
GLUCKMAN FRS
University of Auckland,
New Zealand

CAMBRIDGE
UNIVERSITY PRESS

University Printing House, Cambridge CB2 8BS, United Kingdom

One Liberty Plaza, 20th Floor, New York, NY 10006, USA

477 Williamstown Road, Port Melbourne, VIC 3207, Australia

314-321, 3rd Floor, Plot 3, Splendor Forum, Jasola District Centre, New Delhi - 110025, India

79 Anson Road, #06-04/06, Singapore 079906

Cambridge University Press is part of the University of Cambridge.

It furthers the University's mission by disseminating knowledge in the pursuit of education, learning and research at the highest international levels of excellence.

www.cambridge.org
Information on this title: www.cambridge.org/9780521736206

First published 2011

A catalogue record for this publication is available from the British Library

Library of Congress Cataloging in Publication data
Bateson, Patrick (Paul Patrick Gordon), 1938–
   Plasticity, Robustness, Development, and Evolution / Patrick Bateson,
Peter Gluckman.
   Includes bibliographical references and index.
   ISBN 978-0-521-51629-7 (Hardback) – ISBN 978-0-521-73620-6 (Paperback)
   1. Developmental biology.   2. Evolution (Biology)   3. Phenotypic plasticity.
4. Adaptation (Biology)   5. Survival analysis (Biometry)
I. Gluckman, Peter (Peter D.), 1949–   II. Title.
   QH491.B38 2011
   571.8–dc22

                                                               2010050334

ISBN   978-0-521-51629-7   Hardback
ISBN   978-0-521-73620-6   Paperback

Contents

Preface

'Nobody reads books these days.' Such was the unhelpful advice of one of our colleagues. So why have we written one? Our answer is that, in many areas of biology and psychology, advances in what is known have not been matched by advances in what is understood. This is particularly true when questions about developmental origins are raised in biology and psychology. All too often such debates are reduced to whether a particular feature of an individual is due to nature or nurture. Many scientists in the fields of both biology and psychology have an interest in either development or evolution, or both. Similarly, many practitioners of human and veterinary medicine also wish to have a much clearer understanding of the debate about developmental origins, as do many biological anthropologists and philosophers of biology. This book was written for those who have become dissatisfied with the way in which these debates are conducted.

Biology can show us many wonders, but one of the most remarkable is how a complex organism grows from almost nothing – a tiny seed or a microscopic embryo. Until recently, the processes that are involved seemed beyond understanding and, even now, much remains to be discovered. Nevertheless, the factual certainties have been known for a long time. An egg taken from a hen's nest and incubated artificially does not hatch out as a different species from its mother. Its characteristics were latent from the moment its mother's egg was fertilised by its father's sperm. These *robust* constancies of development are profound and real. At the same time, the hatched chick is capable of adapting to many challenges posed by its environment. It can cope with disabilities generated by accidents or disease. It can learn to recognise particular members of its species and acquire preferences for particular foods that are available to it. The *plasticity* of the bird is as remarkable as its robustness. These twin characteristics, broadly defined, are found throughout the animal and plant kingdoms. They have given rise to a widely held view that the characteristics of any one organism, including the characteristics of a human being, may be grouped into those that are due to the inner workings of its body and those that are due to the external environment.

Along with many others, we feel that it is possible to do a great deal better than that. The treatment needs to be extended to the length of a book because it takes extended argument to change long-established patterns of thought. Words and ideas have to be clarified. In the case of development, the reader needs to have a straightforward view on what is meant by robustness and plasticity – terms that have considerable complexities embedded within them – and then be given an insight into how these seemingly opposed characteristics are integrated in the growing organism. How do the products of such an integration serve the needs of the individual, and how might they have evolved? Finally, have these characteristics of organisms fed back to shape the course of evolution itself?

Our intention in writing this book was to bring clarity to some of the debates that have muddied the waters in biology and psychology for far too long. We have markedly different backgrounds. One of us is a behavioural biologist interested in the development and evolution of whole organisms. The other was trained in paediatrics and has a deep interest in the molecular and evolutionary biology of development. Frankly, we admit that we each sometimes had difficulty with the drafts that the other had written. The molecular stuff was too technical and riddled with jargon, and the behavioural stuff assumed knowledge that is not widely shared by other biologists. We have had to accommodate to each other and recognise that the resulting text might sometimes seem uneven, and maybe irritating, to specialists who have an intimate knowledge of some of the material we present and will judge us to have presented it too simply. However, the issues are so important to biologists and psychologists that the presentation should be cross-disciplinary, and we trust that the overarching picture will make sense. Our hope is that the quality of understanding will start to match the quality of the evidence.

Acknowledgements

A book of this nature very much depends on the many informal interactions we have had over many years with many of our colleagues. More than simply reading the literature, the warm and stimulating interactions with colleagues and students allowed our ideas to develop. We are sure that many of these will see echoes of past conversations, and often intense debate, in what we have written.

Working with some colleagues has been particularly instructive. Mark Hanson of Southampton has worked with both of us for many years and, with P. G., has led a major programme of empirical and conceptual work in developmental epigenetics and its role in human health, and across many of the issues we discuss in this book. Hamish Spencer of Otago has been involved in many discussions with us, and his deep knowledge of the comparative and modelling literature has been invaluable. Kevin Laland who was a colleague of P. B.'s at Cambridge for many years before moving to St Andrews made many helpful comments on the draft manuscript. Eva Jablonka (Tel Aviv), whose work has been seminal in the field covered by this book, also made valuable comments. Mark Blumberg (Iowa), who kindly wrote a pre-publication comment for the book, also went through the entire manuscript and made many helpful suggestions for improvement. We thank these friends and the following from around the world who kindly read parts of the draft manuscript: Alan Beedle (Auckland), Gillian Brown (St Andrews), Tatjana Buklijas (Auckland), Frances Champagne (Columbia), James Curley (Columbia), Partha Dasgupta (Cambridge), Peter Dearden (Otago), John Funder (Melbourne), Lupus von Malzen (London) and Daniel Nettle (Newcastle).

This book would not have been possible without the efforts of Dr Felicia Low of Auckland. She undertook considerable literature searching and reference checking and assisted with the editorial handling of numerous drafts. As a biochemist she had many useful insights on the molecular side of this book, for which we are very grateful.

P. B. is grateful for the generous support of the Leverhulme Trust through the interdisciplinary programme on Human Evolution and Development and P. G. acknowledges the support of the National Research Centre for Growth and Development.

1

Setting the scene

An oak tree growing in a meadow has a relatively short trunk and a familiar structure of wide branching limbs that can easily be recognised. However, when oaks grow in forests they compete with each other and develop long straight trunks; they have an appearance that is totally different from the same species growing in open country. One of us, on a family holiday in Greece, collected two tiny seedlings of the Greek fir from a mountain on the island of Cephalonia. One seedling was planted in a garden in England and, 30 years later, is 15 metres high, growing fast and with a form just like the illustrations in tree books. The second was planted in an earthenware pot and is now a hundredth the size of the other – a perfect bonsai. Such capacity to respond dramatically to the available resources, however limited, and yet survive is also seen in those mammals in which large litters are commonplace. The runt in a litter of pigs might be a tenth of the weight of its siblings at birth, but it is perfectly formed and, if given sufficient milk after birth, will survive to become an adult, albeit of reduced size, that will be capable of breeding.

Nobody will be greatly surprised by these examples, and yet the 'robustness' of development – whereby the general characteristics of each individual develop in much the same way irrespective of the environment – is often contrasted with 'plasticity' or malleability, which allows change, particularly during early development. These seemingly opposed characteristics of organisms are frequently forced into a dichotomy that is often used to explain natural phenomena: the programme for an organism's development is either closed or open; its characteristics are either immutable or subject to change; the brain is either hard-wired or changeable; behaviour is either innate or learnt. We shall argue that these opposing ideas that seem so obvious to many people are misleading and unhelpful to anybody who wishes to understand how the body grows and the brain develops. Our aim is to draw the reader away from patterns of thought that are rooted in conventional public debates about 'nature' and 'nurture' rather than in empirical biology.

The nature/nurture dichotomy is not merely a feature of popular science and folk biology. Many eminent biologists have accepted an 'either/or' account of development. For example in his book *Sociobiology*, Edward O. Wilson (1975) tacitly accepted this position. In a large collection of reviews of the book published in the journal *Animal Behaviour* in 1976, Wilson was attacked because he had not considered the interplay between the developing organism and its environment. In response to these criticisms he wrote that, in his view, development was a black box or 'module' that could be decoupled when the relations between genes and the characteristics of the adult organism were considered (Wilson, 1976). The key message in our book is that development is much more integrated in these relationships than this decoupling image suggests.

The use of the nature/nurture distinction often involved a confusion of categories, since 'nurture' was seen as a developmental process and 'nature' was often viewed as the genetic origin of that process. For some, however, 'nature' was viewed as the adult expression of a developmental process. That point was cleverly captured by Matt Ridley (2003) in the title of his book *Nature via Nurture*. For others, though, nature was reserved for those features that developed 'robustly', unaffected by the vagaries of the environment, and nurture was used for those features that were 'plastic', greatly influenced by the conditions in which the individual developed.

The nature/nurture distinction runs through persistent arguments about the origins of human faculties. The seventeenth-century philosopher John Locke believed that all reason and knowledge was derived from experience. Charles Darwin's cousin, Francis Galton, expressed a strongly contrasting view about the development of human mental faculties, believing that education and environment produce only a small effect on the human mind and that most human qualities are inherited.

The debate continues to the present day. It extends across the full range of human faculties, styles of thinking, and behaviour. The universalists claim that these faculties are shared by and intrinsic to all human beings. The relativists argue that all the cognitive characteristics of humans emerge from the culture in which they are embedded. Reducing the problems of origin to either 'this' or 'that' is deeply unsatisfactory. Equally unhelpful, we shall argue, is the conflation of origins with developmental processes.

These ideas and widely held views about the decoupling of development from evolutionary processes led to constipation of thought and encouraged misunderstanding of how evolutionary theory relates to, and integrates with, ideas about the development of the individual. It has been argued that Darwinian natural selection leads to evolutionary changes in the phenotypic characters of organisms. If these changes occur in stable environmental conditions, it was maintained, they must result from changes in gene frequency. Therefore, or so it was argued, adult characteristics are exclusively under genetic control. Furthermore, understanding development was not essential because Darwinian selection acted on the outcome of each individual's development. Bruce Wallace (1986) expressed a view, shared by many other eminent evolutionary biologists, that an understanding of development was irrelevant to an understanding of evolution. Ron Amundson (2005) has carefully described their position and how it contrasts with the views of those who give prominence to developmental biology.

As the genetic or 'hard heredity' model formed around 1900, other scientists who were focused on the biology of development started to integrate emerging genetic, and later molecular biological, concepts with the developmental framework. Wilhelm Johannsen (1909) separated the organism's 'genotype', or the set of hereditary factors, from its 'phenotype', the organism's developed characteristics. Indeed, he introduced these terms. Richard Wolterek (1909) developed the concept of the 'reaction norm' to denote the range of phenotypes derived from a single genotype under developmental influences. Scientists whose orientation was developmental did not consider development to be immune from the environment. Pioneers in experimental embryology, such as Hans Spemann (1938), identified the role of diffusible factors in the transition from an undifferentiated zygote into a multicellular organism. In the mid-twentieth century Ivan Schmalhausen (1949) and Conrad Waddington (1957), whose work we shall discuss later in this book, provided conceptual and empirical observations that started to show how developmental and evolutionary processes could be reconciled. Richard Lewontin (1983) showed how an organism's activities in changing its environment could affect the evolution of its descendants. In the twenty-first century, influential insights were also provided by Mary Jane West-Eberhard (2003), who suggested – provocatively to some – that phenotypic change was not just a passive follower of genotypic evolution but that phenotypic plasticity could in fact influence evolution, thereby focusing on the importance of developmental phenomena. Similar points were made by the advocates of developmental systems theory (Oyama et al., 2001). Susan Oyama (2000) had already written eloquently on the subject.

Advances in scientific thought in relation to the transactions between the individual organism and its environment are seen especially clearly in the psychological literature (Sameroff, 2010). Encouragingly, some of the advances have passed into the popular domain. David Shenk (2010) has described how each individual interacts with his or her environment in such a way that potentialities may be revealed or suppressed by circumstances. The message is optimistic and has major implications for public policy. Shenk's perspective was presented as an interaction between genes and environment; an image which he drew from the scientific literature (Meaney, 2010). The interaction is often abbreviated as 'G × E'. As we shall outline in later chapters, some problems arise from this formulation because it conflates ideas about sources of variation in populations with those about an individual's development – a point that is powerfully emphasised by Evelyn Fox Keller (2010). It is the person who interacts with the environment, not his or her genes. While genes are definitely activated or repressed by some environmental conditions, the organism can change its environment or choose which environment in which to live without necessarily producing changes in the expression of its genes.

A plethora of ideas and observations have emerged, and we hope to clarify this confusing literature and identify the key concepts. Our overall purpose is to provide an overview of how developmental processes are integrated. We focus on how two superficially opposed sets of processes – those generating robustness and those generating plasticity – operate together in development. However, as we shall explain in Chapters 3 and 4, these are not unitary processes; these apparently simple terms encompass a wide variety of mechanisms operating from the level of the gene to that of the organism.

Undeniably, species are generally recognisable for what they are, and readily seen as distinct from each other. Animals and plants are usually identified correctly by skilled naturalists as members of their own species. Whatever his or her experience, nobody would confuse a human with a rhesus monkey. The general characteristics of each individual develop in much the same way irrespective of the environment in which he or she lives. Many of these features do, indeed, develop robustly in the face of variation in the environment. We shall examine how such robustness comes about at a variety of levels of biological organisation. At the same time, malleability or plasticity during development is a widespread phenomenon in both the plant and animal kingdoms. In many species, a variety of distinct phenotypes may develop from a single genotype, depending on local conditions. Phenotypes may form a continuous range, as in the birth sizes of mammals, or they may be discontinuous, involving quite distinct bodies and behaviour, as in the female honeybee. Plasticity is a term encompassing multiple processes regulated in a variety of different ways. We shall argue that robustness and plasticity are complementary, often integrated in development and therefore difficult to separate. They should not be seen as being in opposition to each other.

Our discussion in this book will encompass both consideration of the whole organism and its mechanistic underpinnings. This means that frequently we will move between different levels of organisation. Many of the same issues about the development and control of the whole organism are also reflected in considerations of its component parts and in the molecular structure and role of the gene. A commonly held view among modern biologists is that primacy and centrality should be given to the gene alone. However, we shall argue that gene expression is profoundly influenced by factors external to the cell nucleus in which reside the molecules making up the genes: the deoxyribonucleic acid (DNA). A willingness to move between the different levels of analysis has become essential for an understanding of development and evolution.

The understanding of the molecular basis of many of the phenomena that we shall discuss is changing rapidly. Increasingly, developmental phenomena can be explained in terms of the differential regulation of gene expression and, in particular, by epigenetic mechanisms that lead to changes in gene expression without a change in nucleotide sequence (see Chapter 5). In some cases these epigenetic effects may be transmitted to the next generation through meiosis, reflecting one of several potential mechanisms of non-genomic inheritance (Gluckman et al., 2007b; Jablonka and Raz, 2009) which must be part of any modern understanding of the essential conjunction of evolutionary and developmental processes.

We shall examine from a biological perspective the interplay between the processes of robustness and plasticity, considering how they evolved and, in turn, how they affect evolution. As active agents in the evolution of their descendants, individuals' mechanisms for responding to change can have profound effects on the rate of evolution and, indeed, on macro-evolutionary processes. We shall go on to consider the particular ways in which these evolutionary processes may work.

Our overall approach arises from the need for a wide variety of disciplines to have a better understanding of developmental processes at this time of immense explosion of empirical data and conceptual understanding. Like many others, we feel that it is crucial to re-integrate development into evolutionary thinking, and to counter the damage done by the emphasis given in so much biological thought to the nature versus nurture dichotomy.

2

Clarifications

MEANINGS OF WORDS

Anyone involved in interdisciplinary research quickly discovers that words do not mean the same to all people. For some, a term is used in the colloquial sense, while for others it has one or more technical meanings. The word 'fitness' is a good example. For the person in the street it refers to physical health and well-being and for the sports physiologist it means something similar, but for the biologist it has a much more technical meaning to do with how likely it is that an individual's characteristics will appear in future generations. Further confusion can be generated by the extensive use of metaphors such as strategy, selection, conflict, design, imprinting, programming and reinforcement, which are borrowed from everyday language but are given technical meanings that sometimes differ from each other across different communities of scientists. In this chapter we attempt to clarify what we mean by the terms that we use throughout this book, and explain why we have chosen not to use some others. In our view this is essential, as much of the literature on development and its relationship to evolutionary process has been confused by misunderstandings that have arisen from alternative usages of the same words. As George Bernard Shaw put it, when talking about the English and the Americans: they are 'two peoples separated by a common language'. Not all our scientific colleagues will agree with our terminology, but at least our meanings will be explicit.

As we have already indicated in Chapter 1, and along with many others, we do not accept the nature/nurture dichotomy, where nature refers to that which is genetically determined and nurture that which is environmentally determined. Inasmuch as we use these terms at all, nature stands for the characteristics of an organism and carries no implication about how they developed. Nurture stands for the processes by which the characteristics develop. In other words, for us the conventional opposition of these two terms is necessarily false.

Later in the book we refer to a number of developmental and evolutionary processes where definitions have been unclear, or where different authors have used different meanings. Such terms include canalisation, stabilising selection, genetic accommodation and genetic assimilation. We shall defer consideration of these terms until subsequent chapters, where clarification of them is appropriate.

ROBUSTNESS

During development, many characteristics of organisms are relatively unaffected by substantial perturbation of the environment and cryptic genetic variation. Their development is, in that sense, 'robust'. Robustness is generally defined as the consistency of the phenotype despite environmental or genetic perturbation (Nijhout, 2002). This implies either insensitivity or resistance to such potential disruption.

We use the term robustness without implying any single or particular mechanism. Indeed we shall argue that many different mechanisms are involved. Robustness is emphatically not an all-or-none phenomenon; it need not affect all systems or organs of the organism in the same way, and it has multiple dimensions reflected in multiple mechanisms. In the next chapter we describe in greater detail the many ways in which robust phenotypes are produced.

PLASTICITY

During their development, individuals with the same genotype may respond to their environments in innumerable and sometimes qualitatively quite distinct ways. We consider plasticity in detail later (Chapter 4) where we describe the overall phenomenon and explain it in terms of multiple unrelated mechanisms. Plasticity includes accommodation to the disruptions of normal development caused by mutation, poisons or accident. Much plasticity is in response to environmental cues, and advantages in terms of survival and reproductive success are likely to arise from the use of such mechanisms. An organism that has been deprived of certain resources necessary for development may be equipped with mechanisms that lead it to sacrifice some of its future reproductive success in order to survive. Plasticity includes preparing individuals for the environments they are likely to encounter in the future on the basis of cues obtained from previous generations; the course of an individual's development may be radically different depending on the nature of these cues. Plasticity may also involve one of the many different forms of learning, ranging from habituation through associative learning to the most complex forms of cognition.

ONTOGENY

The term 'ontogeny' or 'development of the individual' often refers to the processes by which an individual acquires its characteristics, generating what is known as the 'phenotype'. What happens after it has reached adulthood is not included. For us, however, development of a kind continues until death, involving many processes that were also involved early in life. We do not, therefore, distinguish between early learning and adult learning except that the context may make a difference to the outcome, as in behavioural imprinting.[1] Whether or not the underlying processes differ from each other is a matter for empirical study. The stage in the life cycle can be important when defining how resistant or sensitive an organism is in relation to changes in its environment. Furthermore many organisms, humans included, use specialised mechanisms at particular periods in their lives, an example being suckling in mammals. Some mechanisms, such as play by young animals, may be the biological equivalent of the scaffolding used for erecting a building, and are no longer required when the job has been accomplished.

GENES

Some people suppose incorrectly that the characteristics of an organism are encoded in the genes, in the sense that all the information required for its development is contained in DNA. The notion of genes coding for phenotypic characteristics was always problematic, but its limitations have become increasingly apparent as molecular and biological knowledge has expanded.

---

[1] Imprinting is a word used in quite distinct ways in different scientific fields. In this book we deal with two distinct phenomena: behavioural imprinting, whereby an animal forms a preference for a class of stimuli to which it was exposed in early life; and genomic imprinting, whereby certain genes are expressed only from one of the two parental alleles. Behavioural imprinting is divided up into filial imprinting affecting the social preferences of young animals, and sexual imprinting affecting the sexual preferences of adults. It may also refer to the development of other preferences.

For example, a person carrying two copies of the mutated gene that gives rise to cystic fibrosis is destined for illness and a high likelihood of death in early to mid-adulthood. However, the survival of such a person will be influenced by the specific mutation that they carry (at least 1,000 different mutations have been identified in the gene), the lifestyle they lead, the number of respiratory infections they have had, how they are medically treated and so forth.

Much epidemiological research in recent years has been based on sequencing the entire human genome and looking at mutant alleles that correlate with disease. A surprise of these genome-wide association studies has been that even when large populations are studied and the disease of interest is common, such as diabetes, few significant genetic effects are found and the effects of any one specific polymorphism are generally small. Single-gene effects are unusual and largely restricted to relatively rare diseases such as phenylketonuria or haemophilia (Maher, 2008).

Genes have been defined in many different ways: as units of physiological function, units of recombination, units of mutation, or as units of evolutionary process – when they have sometimes been imbued with 'selfish' intentions in order to help with understanding (Dawkins, 1976). The problem of definition has been made worse as it has become clear that the same strand of DNA may serve in processes that differ in function. Indeed, the same strand of DNA might be transcribed in one direction to serve one function and in the other direction to serve a different function. Griffiths and Stotz (2006) emphasise how, in the post-genomic era, the emerging concepts of the gene pose a significant challenge to conventional assumptions about the relationship between genome structure and function, and between genotype and phenotype.

The word 'gene' never had a clear unambiguous meaning: for some it meant simply a sequence of DNA, for others it referred specifically to those segments of DNA that are transcribed into ribonucleic acid (RNA) and then translated into a protein. To be set against that, some segments of RNA – the so-called non-coding RNAs – have regulatory functions, and the term 'gene' is extended by many molecular geneticists to include the DNA sequences coding for these RNAs. These different meanings of gene get conflated, with subsequent confusion of thought (Keller, 2000). As the philosopher of science Lenny Moss has put it with respect to genetic determinism: 'The idiom of the language-of-the-gene became written not by those whose hypotheses were successful but rather by those whose metaphors were successful' (Moss,

2002). Without using any metaphors, we shall simply refer to a gene as a particular sequence of DNA that is inherited from one generation to the next and has functional significance.

EPIGENETICS

Epigenetics is a term that had multiple meanings since it was first coined by Conrad Waddington (1957). He used the term, in the absence of molecular understanding, to describe processes by which the inherited genotype could be influenced during development to produce a range of phenotypes. He distinguished 'epigenetics' from the eighteenth-century term 'epigenesis', which had been used to oppose the preform-ationist notion that all the characteristics of the adult were preformed in the embryo.

More recently epigenetics has become defined as the molecular processes by which traits – as defined by a given gene expression profile – can persist across mitotic cell division without involving changes in the nucleotide sequence of the DNA. Epigenetic processes result in the silencing or activation of gene expression through such modification of DNA and its associated RNAs and proteins. We shall use this definition.

Some authors use a broader definition of epigenetics to include elements apart from DNA, RNA and chromatin, such as components of the cytoplasm outside the cell nucleus and the cell membrane which are passed across in mitosis. In all these usages, epigenetics usually refers to what happens within an individual developing organism.

A growing body of evidence suggests that epigenetic traits established in one generation may be passed directly or indirectly through meiosis to the next, involving a variety of different processes (Gluckman et al., 2007b; Jablonka and Raz, 2009). This evidence will be described in greater detail in later chapters.

FITNESS

In circles of evolutionary biologists, fitness refers to the probability that a trait of an organism will be transmitted to future generations. At its simplest, this is assessed in terms of the likelihood of survival to repro-ductive age of the organism carrying that trait. It can be obtained more accurately from the number of grand-offspring of the grandparent carrying the trait. Grand-offspring are measured because offspring resulting from hybridisation between different species may be infertile.

However, as Bill Hamilton (1964) famously tested by modelling, a trait may be transmitted from one generation to the next by the aid given to a relative other than an offspring. The closer the genetic relationship, the higher the chance that onward transmission will occur when set against the costs of giving aid. Hamilton referred to the overall effect as 'inclusive fitness'. Some theoreticians refer to fitness as the probability that a given genotype will be transmitted from one generation to the next. Consistent with the overall thrust of this book, we shall not use this formulation because it limits understanding of the number of ways in which beneficial phenotypic traits can be transmitted across generations.

INNATENESS

The common usages of 'innateness' are reflected in non-scientific dictionary definitions. These typically refer to innate characters as being present at birth (inborn or congenital), part of the essence of an individual, and not learnt. In the scientific literature 'innateness' has been used in multiple ways and, moreover, is often imprecisely defined by appeals to the role of genes in development, a concept reinforced by the folk concept of 'nature' (Linquist et al., 2011). In turn this reflects widely held views of how Darwinian evolution shapes developmental robustness and the consequent universality of a character within a given species. Blumberg (2005) provided an excellent and highly readable critique of the way in which the closely related term 'instinct' has been used.

In general, the link between genes and innateness is based on imprecise ways of thinking about the role of genes in development (Bateson and Mameli, 2007). No phenotype is such that only genes are needed for its development in the sense that they could, like Japanese flowers, be dropped into water and open up. An alternative formulation that a trait is innate if, and only if, it is genetically influenced is equally unsatisfactory, but for the opposite reason. All phenotypes are genetically influenced because genes participate (one way or another) in the development of all phenotypes, including those that are acquired by learning. It could be argued that innate traits are influenced *distinctively* by genes in ways that non-innate traits are not. The problem is then to specify what this distinctive way might be, and this is not trivial. Part of the difficulty is that environmental factors that are constant in a given set of conditions may nonetheless be important in determining the precise characteristics of a phenotype. In short, linking genes with behaviour is an unsatisfactory account of the role played by genes in development.

The various definitions of innateness raise serious problems when used for all the phenomena that they are meant to encompass (Mameli and Bateson, 2006, 2011). The different usages can also easily lead to the mistaken conclusion that evidence for one meaning of innateness is evidence for another view. This popular use of the innate/acquired distinction serves to generate the illusion that certain important questions have already been answered, while in reality they have not. Hereafter, we shall not use the term 'innate' or the associated term of 'instinct' except when quoting others.

HERITABILITY

Inheritance is central to the Darwinian theory of evolution. The concept of heritability is, however, applied to populations in a variety of ways. In its most straightforward and colloquial sense, high heritability means that closely related individuals have characteristics in common. In evolutionary biology, high heritability is applied to those traits that change rapidly in the course of evolution when driven by natural selection. In population genetics, high heritability means that the additive phenotypic variation due to the apparent effect of genetic variation is high. In this narrow sense, the measure is indicative of the extent to which the variation in the population is due to allelic variation, which is considered independently of interactions with other genes at the same or at different loci. More broadly, heritability refers to the ratio of the spectrum of differences in a characteristic due to genetic variation to the total spectrum of the phenotypic trait in the population. None of these definitions refers to the characteristics of an individual (see Keller, 2010).

A trait has high 'broad sense' heritability in a population to the extent that the existing variation for that trait in the population is due to genetic variation. If variance in a trait is entirely due to genetic variance, broad sense heritability is 1.0; if it is entirely due to variance in non-genetic factors, broad sense heritability is 0.0. Behind the deceptively plausible ratios lurk some fundamental problems. For a start, the heritability of any given characteristic is not a fixed and absolute quantity. Its value depends on a number of factors, such as the particular population of individuals that has been sampled. For instance, if height is measured only among people from affluent backgrounds, then the total variation in height will be much smaller than if the sample also includes people who are small because they have been undernourished. The heritability of height will consequently be larger in a population of exclusively well-nourished people than it would be among people drawn from a wider range of environments. Conversely, if the heritability of height is based on a population with less genetic variation – say, native Icelanders – then the figure will be lower than if the population is genetically more heterogeneous, for example if it includes both Icelanders and African pygmies. Thus, attempts to measure the relative contributions of genes and environment to a particular characteristic are highly dependent on who is measured and the conditions under which they are measured.

Another problem with heritability is that it says nothing about the ways in which genes and environment contribute to the biological processes of development. This point becomes obvious when considering the heritability of a characteristic such as 'walking on two legs'. Humans walk on fewer than two legs only as a result of environmental influences such as war wounds, traffic accidents, disease, or exposure to toxins before birth. In other words, all the variation within the human population results from environmental influences, and consequently the heritability of 'walking on two legs' is zero. And yet walking on two legs is clearly a fundamental property of being human, and is one of the more obvious biological differences between humans and other great apes such as chimpanzees or gorillas. It obviously depends heavily on genes, despite having an estimated heritability of zero. A low heritability clearly does not mean that the trait is unaffected by genes.

The most serious weakness of heritability estimates is that they rest on the assumption that genetic and environmental influences are independent of one another and do not interact. The calculation of heritability assumes that the genetic and environmental contributions can simply be added together to obtain the total variation. In many cases this assumption is clearly wrong, but when techniques are used to calculate the interaction between the sources of variation and an interaction is found, an overall estimate of heritability has no meaning because the effects of the genes and the environment do not simply add together to produce the combined result. For these reasons we shall not use measures of broad sense heritability elsewhere in this book.

ADAPTATION

The term 'adaptation' is used in at least four senses. Two have no evolutionary implications and are used to describe the waning of the responsiveness of a sense organ to a constant stimulus or the response to a changed environment such as an increased rate of breathing in response to a shortage of oxygen. Two other meanings are used in the developmental and evolutionary literature and are often conflated. The first, which we shall use, is for having a feature that is well matched to meet a requirement set by the environment. The second meaning is for the process by which that state is achieved. This might be the evolutionary process of Darwinian selection acting on the trait's effectiveness in a particular context. Alternatively it might be the individual acquiring the trait during its own lifetime through plastic processes. We do not use this last meaning because a state might be achieved in many different ways, including by learning. Therefore, we shall not use adaptation for the process that gave rise to the current state.

We shall use the adjective 'adaptive' to apply to a particular adaptation, defined as a state, and 'adaptability' to describe the capacity of an individual to meet a new challenge set by the environment. The presumption is that adaptations and adaptive responses usually increase the chances of survival and reproduction of individuals expressing them over individuals that do not.

Gould and Vrba (1982) coined the term 'exaptation' for the evolutionary process by which an adaptation is co-opted to meet a new requirement set by the environment. That means that a trait might have a different function in the present from its previous use. For example, the middle-ear bones of the mammal are used to transmit acoustic signals to the nervous system, but they evolved from the jawbone of the fish, where they were used for feeding.

DESIGN

'Biology is the study of complicated [animate] things that give the appearance of having been designed for a purpose', wrote Richard Dawkins (1986) in *The Blind Watchmaker*. Dawkins took the image of the watchmaker from an argument developed by William Paley (1802) in the early nineteenth century. 'It is the suitableness of these parts to one another; first, in the succession and order in which they act; and, secondly, with a view to the effect finally produced', wrote Paley about the reaction of someone who contemplates the construction of a well-designed object such as a watch. Paley supposed that such characteristics, when found in the natural world, must have been created by an intelligent designer. In contrast, Dawkins, following Darwin, believed that seemingly well-designed characteristics were the product of natural selection.

The perception that a feature of an organism is designed stems from the relations between the properties of the feature, the circumstances in which it is expressed, and the resulting consequences. The closeness of the perceived match between the tool and the job for which it is required is relative. In the context of technological design, the best that one person can do will be bettered by somebody with superior technology. If you were on a picnic with a bottle of wine plugged with a cork but with no corkscrew to remove the cork, one of your companions might use a strong stick to push the cork into the bottle. If you had never seen this done before, you might be impressed by the choice of a rigid tool small enough to get inside the neck of the bottle. Tools that are better adapted to the job of removing corks from wine bottles are available, of course, and an astonishing array of devices have been invented. One ingenious solution involved a pump and a hollow needle with a hole near the pointed end; the needle was pushed through the cork and air was pumped into the bottle, forcing the cork out. Sometimes, however, the bottle exploded, and this tool quickly became obsolete. As with human tools, what is perceived as good biological design may be superseded by an even better design, or the same solution may be achieved in different ways.

Although 'natural design' is readily used by many biologists, it has been contaminated by the growth of a new version of creationism – namely intelligent design – the sense in which William Paley used biological adaptations as evidence for the existence of God. Therefore, to avoid confusion, we shall not use the metaphor of 'design' any further in this book.

DETERMINATION AND CAUSALITY

The concept of 'determination' creates difficulties in biology. Just because a mutation in a gene is correlated with a phenotypic change in a population study, it does not mean that that particular gene is the sole or most important determinant of the trait, or even that it is directly causal at all. The fallacy of association being conflated with causation applies as much to genome-wide association studies as to other areas of science. Just because a gene is knocked out in a commonly used experimental animal, such as a mouse, and causes no obvious phenotypic change in one set of developmental conditions does not mean that it will not influence phenotypic development under a different set of experimental conditions.

Unfortunately, however, the language of genetic determinism dominates much of biology. Rarely, if ever, does any biological phenomenon have a single cause and, for that reason, such deterministic language is misleading. Outcomes for organisms are influenced by a great many factors. The environmental ones, which involve the individual's development, were often tucked away into a black box for convenience by most evolutionary biologists. The black box is now being opened to provide a more complete picture of what really happens.

Given that a large number of factors affect development, we note that a variety of qualitatively distinct environmental factors influence how development proceeds. Some may enable the expression of a broad suite of characteristics and a particular integrated phenotype (a morph), while others may influence the development of a limited set of characters. For example, temperature-dependent sexual differentiation in sea turtles affects many anatomical, hormonal and behavioural features (Davenport, 1997). Temperature exposure in the human infant, if it has any effect at all, is highly specific and may only affect the innervation of dermal sweat glands (Diamond, 1991).

Finally, the notion of one cause leading to one effect, implicit in the idea of chain-like determination, comes unstuck when feedback mechanisms are involved, as is often the case in biology. To take a commonplace example from everyday life: does room temperature determine when a heating system is turned on or off by a thermostat, or does the heat generated by the heater determine the temperature of the room? These issues become important when discussion moves outside the controlled environment of the laboratory into the free-living world in which organisms exist.

DIFFERENT PROBLEMS

Organisms and their component parts are studied in many ways because people become absorbed in different types of problem. Some want to know, for instance, how a particular character or system benefits the organism, while others want to know how it works. Clearly a number of fundamentally different types of questions may be asked when studying biology. Probably the most useful and widely accepted classification was formulated by the Nobel Prize-winning ethologist, Niko Tinbergen. He pointed out that four distinct types of problem are raised in biology: mechanism, ontogeny, function and evolution (Tinbergen, 1963). These then relate to four different questions that can be asked about any feature of an organism: 'How does it work?',

	Current	Historical
Proximate	How does it work?	How did it develop?
Ultimate	What is it for?	How did it evolve?

Figure 2.1. Tinbergen (1963) posed four distinct questions that may be addressed when examining a biological or psychological phenomenon. Questions about mechanism and biological function deal with the present. Questions about evolution and development deal with the past. The mechanistic and developmental questions are sometimes called proximate and those about function and evolution are sometimes called ultimate. From Martin and Bateson (2007).

'How did it develop?', 'What is it for?' and 'How did it evolve?' Questions of current utility and evolution are often referred to as 'ultimate' questions, in contrast to mechanism and development. In the case of a fully formed feature of an organism, questions to do with mechanism and function are current, whereas questions to with evolution and development are historical (see Figure 2.1).

Tinbergen's distinctions can perhaps best be illustrated with a commonplace example. Suppose we ask why it is that drivers stop their cars at red traffic lights. One answer would be that a specific visual stimulus – the red light – is perceived, processed in the central nervous system, and reliably elicits a specific response (easing off on the accelerator, applying the brakes, and so on). This would be an explanation in terms of mechanism. A different – but equally valid – answer is that individual drivers have learnt this rule by past observation and instruction. This is an explanation in terms of development (ontogeny). A functional explanation is that drivers who do not stop at red traffic lights are liable to have an accident or, at least, be stopped by the police. Finally, an 'evolutionary' explanation would deal with the historical processes whereby red came to be recognised as representing danger and a red light came to be used as a universal signal for stopping traffic at road junctions. All four answers are equally correct, but reflect four distinct levels of enquiry about the same phenomenon. This approach to biological phenomena, together with appreciation of the different intent of each of these four questions, is a valuable tool when approaching the issues that we consider in this book.

In particular, Tinbergen's questions about current function and past evolution highlight an essential, but often ignored, distinction to which we shall return. A trait may have initially evolved to be of benefit in one situation but may become useless or may evolve (by exaptation) an entirely different function in another environment. Some traits may not be adaptations at all, but rather are the incidental consequence of Darwinian selection on other traits – for which Gould and Lewontin (1979) adopted from architecture the term 'spandrels' in their famous essay.

The general point about the distinctions made by Tinbergen and repeated in almost every modern textbook about animal behaviour is the valuable clarifications they brought to discussions among both biologists and psychologists. Function and mechanism were often confused. Consequently, when one scientist had been considering the current utility of a trait, another might have thought the discussion was about the mechanisms that underlie that trait. Such misunderstandings can be avoided by appreciating that, unsurprisingly, different perspectives raise different questions. Once clarified, the different approaches work to help each other, and the bringing together of different approaches is something that we pursue in this book.

3

Developmental robustness

When Carl Linnaeus developed his classification of living organisms in the eighteenth century, he was probably clear in his own mind just what constituted a species. Each one was clearly distinct physically and could be recognised as such. It was God-given. Even when the development of evolutionary theory took off with the publication of Charles Darwin's *On the Origin of Species*, biologists continued to use the familiar Linnean binomial of genus and species for each organism. This was because species recognition was assumed by many to be straightforward and because of the presumed continuity over time from distinct ancestor to distinct descendant. Such views were challenged at the time, most notably by the French biologist Jean-Baptiste Lamarck, and in modern times the definition of a species is a source of much controversy among biological theorists. Nevertheless, the readily recognised features of a given species are generally familiar to anyone who has used one of the innumerable field guides or botanical keys. Such constancies raise an issue which is central to this book. Something about a house sparrow ensures that no member of its species becomes a crow. Gross atypical morphologies occur, of course, but they are usually dysfunctional and in the past were termed 'monsters'. Many structural attributes of an organism, such as the number of limbs or digits, are invariant and the molecular basis of this consistency of developmental pattern is increasingly understood.

The resistance of bodies to deviation from the form or forms that are typical for the species is also expressed in behaviour. The views of the founders of modern ethology, Konrad Lorenz and Niko Tinbergen, on instinct were based on many compelling observations of animals' behaviour in both captive and natural conditions (Burkhardt, 2005). These have been added to by a great wealth of evidence in subsequent years, and many examples of courtship, defensive behaviour, specialised feeding methods, communication and much else have become familiar to a wide audience through remarkable television films. Some behaviour patterns are highly stereotyped in their form and are stable across a wide range of environmental conditions.

Complex and coordinated behaviour patterns may also develop without practice. Birds, for example, can usually fly at their first attempt, without any apparently relevant prior experience. In one experiment, young pigeons were reared in narrow boxes that physically prevented them from moving their wings after hatching. They were then released at the age at which pigeons normally start to fly. Despite having had no prior opportunity to move their wings, the pigeons were immediately able to fly when released, doing so almost as well as the pigeons that had not been constrained (Grohmann, 1939). In a similar way, European garden warblers that had been hand-reared in cages nevertheless became restless and attempted to fly south in the autumn – the time when they would normally migrate southwards. The warblers continued to be restless in their cages for a couple of months, which is equivalent to the time taken to fly from Europe to their wintering grounds in Africa. The following spring, they attempted to fly north again. This migratory response occurs despite the fact that the birds have been reared in social isolation, with no opportunities to learn how to migrate (Gwinner, 1996).

The comparative approaches to classifying organisms have emphasised the similarities in the characteristics of members of the same taxonomic group. These approaches often rely on molecular techniques that usually confirm, but occasionally contradict, classifications based on gross morphology or behaviour. In this chapter we outline the ways in which members of each species retain their identity despite great differences in the developmental environment or the way in which they have been reared.

Beyond species-level consistency, the constancy of an individual's characteristics is maintained in the face of both environmental and genetic variation. The developmental biologist Frederik Nijhout has proposed a formal approach to robustness in which he suggests that the developing organism is robust if it is unable to detect changes in the environment or is resistant to them (Nijhout, 2002). Such resistance could also provide the basis of robustness to genetic perturbation. While this was a critical contribution, insensitivity or resistance to perturbation does not encompass all the processes by which consistency of phenotype is attained. In this chapter we take a broader view of the variety of ways in which robust phenotypes can be generated. What follows is a classification of likely mechanisms.

INSENSITIVITY

If a developing organism cannot detect an environmental change, then it cannot respond to it. This was part of Nijhout's definition of robustness. The organism may not have the sensory equipment that is sensitive to change, or a barrier may exist between itself and the change. For example, the dependence of the bird embryo on its yolk sac for nutrition means that it is insensitive to changes in the nutritional environment of the mother after the yolk sac has been formed. In that respect the bird embryo differs from the mammalian fetus, which is sensitive to some aspects of the nutritional environment of its mother throughout gestation.

In mammals the placenta can modulate or act as a barrier to a number of potential environmental changes. The fetus has no need to thermoregulate, which is energetically expensive; instead the mammalian fetus is held at a relatively constant temperature which is slightly higher than its mother's because the placenta secretes factors into the fetal circulation which inhibit the burning of brown fat. At birth, thermogenesis is induced by the severance of the umbilical cord (Gunn and Gluckman, 1989). In other words, the fetus is effectively insensitive to its thermal environment – it can neither increase heat production nor reduce heat loss. However, this robustness has a cost because the fetus is passively dependent on maternal temperature regulation, so if the mother develops a fever the fetal temperature will also rise. This may have adverse consequences; for example, the hyperthermic brain is more likely to suffer brain injury from asphyxia, and thus maternal fever during labour is a risk factor for perinatal brain injury associated with birth asphyxia.

Many potential toxins cannot cross the placenta but others can. Thalidomide does cross the placenta and one of its metabolites – which is toxic – leads to the disruption of fetal development associated with maternal thalidomide use. Maternal cortisol levels fluctuate considerably in response to stress or exercise, but the placenta contains an enzyme, 11-beta-hydroxysteroid dehydrogenase type 2, that can convert the active steroid into an inactive form (Brown et al., 1993). Consequently, the fetus is relatively unaffected by changes in maternal cortisol levels except under extreme conditions. In other words, with respect to fluctuations in maternal stress, the development of the fetus is relatively robust.

CONSTRAINTS

A multicellular organism cannot maintain infinite plasticity in all its attributes. Stability of organisation requires a series of irreversible steps; for example, the differentiation of embryonic stem cells into particular cell types is a fundamental aspect of development. Once committed to differentiation, cell lineages do not normally de-differentiate. Rapidly dividing cancer cells were thought to be an exception to this rule, but most cancers arise from latent stem cells rather than de-differentiation of differentiated cells (Rosen and Jordan, 2009). In most sexually reproducing species, sexual differentiation is irreversible and sets a constraint on what can subsequently occur; some fish such as the wrasse are notable exceptions.

As a result, some changes in form and state may simply not be possible. Some constraints are imposed temporally by what has happened earlier in development and the architecture of the underlying molecular networks (Arthur, 1997). The analogy to a building is obvious: just as it is impossible to redesign the foundations once a skyscraper is almost complete, once the basic organisation of an organism is established, it cannot be revisited. Even when the organism changes its form though metamorphosis, the basic generalised body plan is maintained. The implications of such developmental constraints, and the impact of adverse events occurring in sensitive periods, are discussed further in Chapter 5.

Other constraints are imposed by the provision of limiting factors such as nutrients. For example, the mammalian fetus is constrained in its growth by placental function and, in particular, nutrient and oxygen delivery. Limitations may be imposed by the placental clearance of fetal growth-promoting hormones. These mechanisms are sometimes grouped together and termed processes of maternal constraint (Gluckman and Hanson, 2005). As a result of these processes, the mammalian fetus generally does not grow at a maximal rate; rather, fetal growth is coordinated with maternal size, which markedly reduces the risk of maternal and infant mortality.

Maternal constraint can be demonstrated by embryo transfer experiments. If an embryo is transplanted into a larger uterus, the fetus will grow larger, demonstrating the importance of nutrient supply in determining birth size. The classic demonstration of this was the work of Walton and Hammond (1938) studying the outcome of crossbreeding between Shetland ponies and Shire horses. The fetus of a crossbreed grew proportionately to the size of the dam – that is, a Shire–Shetland cross in a Shire uterus grew much larger than the reciprocal Shetland–Shire cross in a Shetland uterus, even though the genetics of each cross were similar. In recent years the non-genetic nature of this phenomenon has been well established in embryo transplant experiments, so that in human pregnancies conceived using donor eggs, birth size is more closely related to the recipient mother's size than to the donor mother's size (Brooks et al., 1995).

ELASTICITY

Elasticity is a term that originally came from the analysis of inanimate materials and was used in contrast to plasticity. It was used to define a structure that can be deformed by a physical force, but once the force is removed, the structure would regain its previous form. For example, a rubber band can be stretched but will then revert to its original size. Some of the resilience or robustness seen in organisms after an initial response to a change in the environment might be explained in similar terms, although other aspects of such resilience, such as wound repair or catch-up growth after starvation or disease, probably involve active regulation, which we consider later in this chapter. The skin is an elastic tissue. Nevertheless, the relevance of the concept of elasticity to development has yet to be established.

ATTRACTORS AND SYSTEMS THEORY

Another conceptual explanation for robustness comes from dynamical systems theory. Certain states are more stable than others, so that in dynamic systems they will be favoured over time, thus acting as attractors. For example, a swinging pendulum will invariably come to rest in the same position – perpendicular to the ground – regardless of the position from which it was released. In biological terms, the characteristics of a system are stabilised by its attractor, and in that sense the system is robust (Thelen, 1989). Attractors draw to themselves phenotypic characteristics that might have ended up in a variety of different places depending on local conditions (Huang, 2009). When formalised mathematically, an attractor can be a point, a curve, or a complicated set known as a 'strange attractor'. The way in which the phenotypic characteristics of an organism are drawn to an attractor does not have to satisfy any special constraints. The process might involve straightforward physical principles, as Stuart

Newman has argued with respect to the formation of segments in the embryo (Newman, 2007).

Usually the explanation for robustness based on attractor theory does not propose a mechanism, but simply states that attractors are a fundamental feature of complex systems. However, Nijhout (2002) demonstrated that feedback loops and dynamical interactions may limit the number of end states that are possible. Such explanations offer another way of thinking about how a specific end state can be achieved in many different ways – that is, the phenomenon of 'equifinality' which excited systems theorists such as von Bertalanffy (1974). To take a biological example, cats can acquire and improve their adult predatory skills via a number of different developmental routes: by playing with their siblings, by playing at catching prey when young, by watching their mother catch live prey, by practising catching live prey when young, or by practising when an adult. Hence a kitten deprived of opportunities for play may still develop into a competent adult predator, but by a different developmental route (Martin and Caro, 1985). The general point is that organisms may reach the same end-point via many different pathways.

REDUNDANCY

In complex machines designed by humans, such as an aeroplane, back-up systems are commonly provided so that if one fails another can be brought into operation. Lives depend upon them. Such redundancy is also found in organisms. The provision of alternative systems protects against failure, and from time to time animals will inevitably be faced with the situation where no amount of tactical manoeuvring will enable one of their developing systems to proceed along a particular route. Such an animal is in a position similar to a human arriving at a railway station only to find that the trains have been cancelled. The traveller can still reach his or her destination but only by choosing a different method of getting there. The idea of equifinality mentioned in the previous section is essentially a statement about redundant mechanisms, without specifying how they work.

Mammals have a number of paired organs that allow for redundancy and back-up in function, such as the kidneys, the lungs, and the parathyroid and adrenal glands. They also have an appreciable reserve within many organ systems. For example, the human kidney has a considerable reserve of nephrons, and the human brain a considerable reserve of neurons. At the biochemical and physiological level many systems have a level of redundancy. For example, both shivering and non-shivering thermogenesis help to maintain body temperature. Human growth hormone has an important role in glucose homeostasis, but growth-hormone-deficient adults and most growth-hormone-deficient children generally have no abnormality of glucose homeostasis unless they have other hormonal abnormalities.

One of the most unexpected aspects of studies in functional genomics has been the number of cases where the removal of a gene by way of recombinant technology, leading to the 'knockout' of a gene in a mouse or other model organism, had no obvious effect on the phenotype. For example, mice lacking the *Hox C* gene cluster, which is known to be involved in body pattern formation, still possess the correct overall body plan (Suemori and Noguchi, 2000).

Redundancy is common at the molecular level. Many genes are duplicated, particularly in vertebrate evolution, less so in invertebrates. Gene duplication is a common phenomenon in molecular evolution, and up to 5% of the human genome consists of duplicated segments (Eichler, 2001). Gene duplication is commonly demonstrated in regulatory systems involving hormones or cytokines and their receptors. In other cases the duplicated genes may show redundancy of function. On the long arm of human chromosome 17 is a cluster of four genes and one pseudogene (a duplicated but non-functional gene) that are expressed in the placenta and are involved in the production of growth hormones that regulate maternal metabolism in favour of fetal growth. Normally only two of the four gene products are expressed in the placenta, but other members of the cluster may be able to compensate after partial gene deletion, and the resultant fetal development is normal (Fuglsang and Ovesen, 2006).

REPAIR

Many organ systems are capable of a degree of robustness through repair, which manifests as hypertrophy (e.g. heart muscle), hyperplasia (e.g. skin, bone) or activation of stem cells (e.g. bone marrow). In vertebrates, one of the most dramatic examples is that of limb regeneration in the salamander. Uniquely, these animals can fully rebuild a lost limb even in adulthood. In contrast, other amphibians such as frogs may be able to do so in the larval form but not in the mature form. Some evidence suggests regenerative capacity even in the mammalian embryo (Stocum, 2006).

REGULATION

The well-regulated body has been a hallmark of physiology since Claude Bernard's writing about the maintenance of *le milieu intérieur* in the nineteenth century and Walter Cannon's elegant exposition of homeostasis in the first part of the twentieth century (Cannon, 1929; Holmes, 1963). Stability is an inherent feature of feedback loops and many physiological systems involve direct or indirect feedback, creating a relatively robust state.

The set-point of a physiological feedback loop is central to the concept of homeostasis and provides one form of robustness. These set-points may change across the life course in a process known as 'homeorhesis' in a developmental context (Waddington, 1957), or as 'rheostasis' for all cases where the physiological set-point can change (Mrosovsky, 1990). For example, within the hypothalamic–pituitary–gonadal system of mammals, a period of gonadal activation in fetal life is associated with low sensitivity of the brain to negative feedback by sex steroids – a system known as the gonadostat (Rosenfield et al., 2008). In the male, this control system is necessary for full development of the external genitalia. After birth, the brain becomes sensitive to negative feedback and remains so throughout the juvenile period. It then again becomes less sensitive at puberty. In seasonally reproducing species such as the red deer, a seasonal switch in the gonadostat, driven by day length, moves from insensitive to sensitive feedback as the days get shorter (Clutton-Brock et al., 1982).

External feedback may also involve other individuals. If a goat has twins, the mother will lick the stronger of the twins, pushing it away from her teats and so promoting the weaker one's development. As the initially weaker kid receives more milk and outpaces its sibling's growth, the mother will now lick it more intensively to allow the now less vital sibling to have more milk; this reciprocal feedback between mother and her two kids continues until the kids have viable independence (Klopfer and Klopfer, 1977).

The ways in which the cells of the body differentiate for their specialised function may be specified by particular amounts of soluble substances produced at sites remote from the cell. The regulation of these soluble substances, known as morphogens, is extremely robust (Eldar et al., 2006). Understanding the processes that buffer morphogen gradients against genetic and environmental perturbations is an area of active research. MicroRNA molecules are likely to play an important role in such regulation (Li et al., 2009).

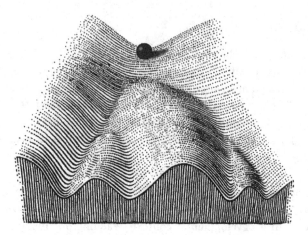

Figure 3.1. The epigenetic landscape as envisaged by Waddington (1957). The ball at the back of the drawing represents an undifferentiated cell. As the ball rolls down one of the valleys, its fate is increasingly determined and constrained as it reaches each choice point in the landscape. Study of the factors that determine the choices and the character of the differentiated cell are all part of the subject that Waddington called epigenetics. From Waddington (1957).

CANALISATION

In the development of an individual, the maturation pattern of organs and behaviour is well regulated. In presenting his image of developmental canalisation, Waddington (1957) offered a visual aid to the biologist who has difficulty in grasping the abstractions of a purely mathematical model. He represented the development of a particular part of an embryo from a fertilised egg as a ball rolling down a tilted plane which is increasingly furrowed by valleys (Figure 3.1).

He called the surface down which the ball rolls the 'epigenetic landscape'. The mounting constraints on the way in which the phenotypic character can develop are depicted by the increasing restriction on the sideways movement of the ball as it rolls towards the front lower edge of the landscape. The landscape therefore represents the mechanisms that regulate development. Waddington's model is attractive to the visually minded because it provides a way of thinking about developmental pathways and the astonishing capacity of the developing system to right itself after a perturbation and return to its former track. The phenotypic character is developmentally 'canalised'. He defined canalisation as 'the capacity [of development] to produce a particular definite end-result in spite of a certain variability both in the initial situation from which development starts and in the conditions met during its course' (Waddington, 1957). A similar idea was formulated by the Russian biologist, Schmalhausen (1949). At times the term canalisation has been used to address multiple processes of conferring robustness (e.g. Flatt, 2005), but we restrict our usage to the phenomena that Waddington described.

Attempts to define the molecular basis of such developmental canalisation have focused on the buffering role of the heat shock protein in the fruit fly (see Chapter 5), but doubtless other mechanisms will come to light. Indeed, secondary enhancers may prove to be a case in point. Enhancers are regions of DNA that do not code for proteins and, when activated, promote gene expression; the primary ones are close to the target gene, the secondary ones are more remote (Perry et al., 2010). In *Drosophila* the secondary enhancers appear to buffer the genome against perturbation, thus providing a canalising process (Frankel et al., 2010).

DIFFERENTIAL SURVIVAL OF SYSTEMS AND COMPONENTS

Overlapping with the concepts of redundancy and regulation in providing mechanisms of robustness is that of developmental selection. Throughout the process of forming the body and the brain, and hence behaviour, the characteristics that develop are the ones that work best as an integrated whole. According to this view, integration of the whole body involves differential survival of particular functional systems and their sub-components. At the cellular level, this involves the well-known process of regulated cell death or apoptosis (Danial and Korsmeyer, 2004). A normal component of brain growth involves the formation of many more neurons than are required. Only those neurons that are attracted to others, and that form connections appropriately and establish functional networks are maintained. Cell death and neuronal pruning are well under way from before birth in species such as the human. Use modifies the rate and pattern of loss and confers a robustness of function.

CONCLUSIONS

This chapter described the multiple levels and some of the processes that lead to a relatively invariant outcome. Although robustness is often defined in relation to insensitivity or resistance of the phenotype to environmental perturbations, many of the same principles apply to perturbations in the genome. Elaborate cellular mechanisms maintain the fidelity of DNA replication during mitosis and of recombination during meiosis; these processes are further examined in Chapter 7. Even more robustness is provided by diploidy (having two copies of a chromosome), as is evident from the greater rates of mutation in the haploid (single copy) Y chromosome, haploid organisms such as the male honey bee and in haploid mitochondrial DNA. DNA repair is important in maintaining constancy of character within a species.

In summary, many different processes, working in different ways, can ensure that members of the same species end up looking alike. These processes can operate at many different levels, ranging from the molecular to the behavioural. Whatever their nature, robustness at one stage of development does not necessarily imply robustness at another. A trait that is robust up to one stage may not continue to be robust thereafter, and vice versa. Developmental malleability may be followed by non-malleability, as in many examples of alternative phenotypes found throughout the animal kingdom, including humans (Gilbert and Epel, 2009). Conversely, developmental non-malleability may be followed by considerable malleability, as in the case of the human smile, which reliably appears in infants during the fifth or sixth week after birth and is subsequently greatly modified by social interactions and cultural influences (Bateson and Martin, 1999). Robustness is not necessarily an all-or-nothing phenomenon, need not affect all systems or organs of the organism in the same way, and has multiple dimensions reflected in multiple mechanisms.

4

Plasticity

The term 'plasticity' refers to the changeable character of matter. It is used in physics for inanimate materials and there it is contrasted with 'elasticity'. If a coiled spring is pulled beyond the limits of elasticity, it will be permanently elongated. Provided that the spring does not break, the change is plastic. In the nineteenth century, the term was introduced into medicine to refer to the renewal of injured tissue and into popular literature to refer to impressionable minds. Plasticity was a dominant theme of James Mark Baldwin's (1902) book. It has returned in many other works about behaviour and the nervous system (e.g. Horn et al., 1973; Gollin, 1981; Lerner, 1984; Rauschecker and Marler, 1987).

Nowadays, plasticity, defined broadly in terms of malleability (see Pigliucci, 2001), is applied across a broad range of biological phenomena, and this extensive usage can cause confusion if the particular use is not well defined. Muscles that are not used diminish in size (atrophy) and those that are exercised become larger (hypertrophy). These are reversible phenomena. Many other cases occurring early in development usually are not. When one kidney fails to form, the other kidney undergoes compensatory hypertrophy and the outcome is stable. The behavioural repertoire of an individual can be changed by one of the many different types of learning, and at the molecular level the immune system responds to infection by developing a long-lasting reaction to the specific virus or parasite that caused the infection. In social insects, the particular way in which the larvae are nourished establishes whether the individual devotes her life to reproduction or to caring for the colonial nest.

Such induced qualitative differences (that is, discontinuities in the range of phenotypes induced) that are found between members of the same species are sometimes referred to as 'polyphenisms' (Mayr, 1963) or morphs. Alternatively the plastic response may induce quantitative differences across a continuous range and these are referred to as 'reaction norms' (Wolterek, 1909; Schlichting and Pigliucci, 1998).

In this chapter we describe these and the other examples of phenotypic variation affected by developmental processes, to give a flavour of plasticity in the natural world, both at the level of the whole organism and at lower levels of organisation. The examples include coping with disruption of normal development, different phenotypic outcomes generated by different cues early in development, learning, and the plasticities of the nervous and the immune systems. The key question is whether these vastly heterogeneous phenomena have anything in common with each other. Whether or not they are necessarily or even plausibly related, they are all of great biological significance.

ACCOMMODATION

An individual whose body has been damaged in an accident or who is burdened with a mutation that renders its body radically different from other individuals may be able to accommodate to such abnormality (West-Eberhard, 2003). In doing so the individual may develop novel structures and behaviour not seen in other individuals of the same species. Such accommodation can be particularly marked when it occurs early in development. An oft-cited example is that of a goat born without forelimbs that walked about on its hind legs and developed a peculiar musculature and skeleton (Slijper, 1942; West-Eberhard, 2003). The coping ability shown by the bipedal goat and the resulting effects on its behaviour and skeleton illustrate one form of plasticity that is termed phenotypic accommodation. The organism has coped with an abnormality by accommodating to it. Similarly, humans born with limb abnormalities as a result of exposure to a teratogen such as thalidomide develop strategies to cope, for example by moving objects using their feet or teeth in ways that others might use their hands for (von Moltke and Olbing, 1989).

The ability of adult humans to deal with serious injury through accommodation can be quite remarkable. One case was that of Jesse Sullivan, who worked on high-voltage electricity lines as a technician for a power company. When in his 50s, he was badly electrocuted. Both his arms were terribly damaged and had to be amputated. In due course he had prosthetic arms fitted. One of these, the left arm, was a marvel of mechanical and electrical engineering. Electrodes from the artificial arm were attached to his chest muscles and within six months he was

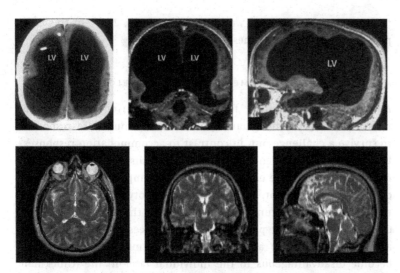

Figure 4.1. (Upper panel) Scan of the brain of a man with enormously enlarged ventricles and very thin cerebral cortex. Despite the abnormality, the man is behaviourally and socially normal. LV = lateral ventricle. From Feuillet et al. (2007), reprinted with permission from Elsevier. (Lower panel) Brain scan of a normal person. Image courtesy of K. A. Johnson and J. A. Becker, from *The Whole Brain Atlas* at www.med.harvard.edu/AANLIB/home.htm.

able to control the movements of the artificial left hand (McGrath, 2007). Such was his ability to accommodate that he was able to do something quite new and different with muscles that had never been required for such tasks during human evolution.

In both of these examples, the compensatory response may involve either behavioural or structural change, or both. In humans who are subject to neurological injury, considerable accommodation may occur, particularly if the injury occurs early in life. Some children with brain injury *in utero* or at birth may be functionally relatively normal as a result of co-option of other neural elements. For example, if a child is born blind, the occipital cortex normally associated with vision may be co-opted for auditory and tactile purposes. In adults who have strokes that lead to the death of neurons, over time the impact of the injury may lessen as new neural connections are formed and other neurons are co-opted to serve the lost functions. The remarkable reorganisation that can take place is shown in a brain scan of a person who had hydrocephalus as a young child (Figure 4.1). The cerebral cortex is extremely thin, yet the person is behaviourally normal and functions as a white-collar worker (Feuillet et al., 2007). In this case, plasticity generated robustness of function.

IMMEDIATE ADJUSTMENTS IN EARLY DEVELOPMENT

Another form of 'coping', found especially during early development, arises when the organism must make immediate responses to survive a challenge but, in contrast to accommodation responses, the normal developmental sequence is not disrupted. While these responses may involve either structural or temporal changes in the course of development, in contrast to phenotypic accommodation they do not entail a fundamental change in the normal pattern of development. Thus the phenotypic consequences are not as dramatic as those that involve accommodation, but may have a cost and become disadvantageous to the individual later in life (Gluckman et al., 2005b).

The timing of metamorphosis in the spadefoot toad (*Scaphiopus hammondii*) is accelerated when the desert ponds in which the tadpoles live start to dry up. In response to the evaporation (detected as an increase in population density), the tadpoles undergo earlier metamorphosis and as a result grow into small adults which are more subject to predation as a result of their smaller size (Denver et al., 1998). Similarly in mammals, if the mother is undernourished or if the placenta is not delivering optimal nutrition, the offspring may be born smaller than usual, with the consequences of greater infant mortality and lower fitness in later life resulting from persistent growth failure (Gluckman et al., 2005b). In polygynous species, such as red deer, the fitness costs may be severe because a small male is less able to compete with larger males for mates and, as a consequence, has a much lower chance of fathering offspring. Nevertheless, survival means that the small male does have some possibilities for mating unobtrusively when larger males are not looking (Albon et al., 1987; Kruuk et al., 1999). In humans, growth retardation following placental insufficiency may be associated with reduced muscle mass, bone density, adult size, and cognitive and attentive function. These neurological effects may be related, by extrapolation from experimental studies, to a trade-off between investing for the long term in neural capacity and the need to expend the limited energetic supply for immediate survival (Gluckman and Hanson, 2005). We return to consideration of the trade-offs implied here in Chapters 6 and 7.

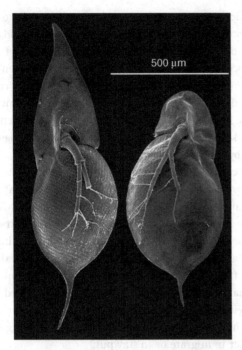

Figure 4.2. Scanning electron micrograph of the small freshwater crustacean *Daphnia* whose mother was exposed to kairomones released from the tissues of other *Daphnia* killed by predators (left) or whose mother was not so exposed (right). The presence of predators induces lengthening of the offspring's helmet to approximately double that of offspring whose mother was not exposed to the kairomones. From Agrawal et al. (1999), adapted by permission from Macmillan Publishers Ltd.

FORECASTING FUTURE ENVIRONMENTAL CONDITIONS

In the cases discussed in the previous section, the organism benefits from an immediate response to a challenge. However, some plastic responses induced in early life may have delayed benefits, so that their primary or only adaptation is expressed at a much later stage in the life cycle. Such anticipatory responses rely on the cue in early life predicting some characteristic of the future environment. The distinction from the previous class of response is not absolute, and advantage from a plastic response may exist both early and late in the life cycle.

Many animals and plants develop defensive structures if they are exposed to cues associated with predators early in life, thus conferring potential advantages in terms of survival if they are later to confront such predators. A classic example, shown in Figure 4.2, is the small freshwater crustacean, *Daphnia*, which develops a defensive helmet and tail spike only if its mother has come into contact with water containing the body fluids of other *Daphnia* that had been eaten by a predatory midge (Laforsch et al., 2006). Many other induced defensive responses to predation are seen in much more complex animals (Gilbert and Epel, 2009). The crucian carp (*Carassius carassius*) found in lakes containing predatory pike have much deeper bodies than those found in lakes without pike, and are consequently more resistant to attack. Experiments showed that this morphological difference was induced by the presence of pike (Brönmark et al., 1999).

The migratory locust (*Locusta migratoria*) can be found as one of two common morphs, with some intermediate forms. If the population density is high, the locust will develop into the migratory morph with large wings, gregarious behaviour and an omnivorous diet. In contrast, when the population density is low, the locust develops into a solitary morph with small wings, reclusive behaviour and a selective diet (Applebaum and Heifetz, 1999). At the larval stage of development, when the trajectory that will determine the adult form is induced by environmental cues, neither form confers a particular advantage under the local conditions. The advantage comes later when, depending on conditions, it pays either to migrate or to stay put.

Two further examples of developmental anticipation will be described in considerable detail because they both demonstrate important points and both have generated a large and active research agenda extending from the whole organism to the molecular level.

MATERNAL STRESS

The study of the impact of early experience on the stress response of the rat has a long history. Levine (1957) reported that handling rat pups caused them to develop in quite different ways from non-handled pups. He subsequently found that when humans came through the animal house, the rats that had been handled in early life were at the front of their cages and the non-handled rats, apparently more frightened, were at the back of their cages (Levine, 1969). Denenberg referred to the effects of the early experience as 'programming' of the rats (Whimbey and Denenberg, 1967), foreshadowing the use of this slightly unfortunate term in the literature on the developmental origins of health and disease (DOHaD).

For many years the non-handled rats were regarded as the control group. However, in a laboratory animal house where food is available all the time, and temperature, humidity and so forth are constant, rat mothers do not care for their pups as much as they would in natural conditions. The deprivation of maternal contact has a major effect on the offspring's behavioural temperament (see Thoman and Levine, 1970). The adverse long-term effects can be prevented if the pups are handled by humans while they are still with their mother. The presumption is that the handled pups emitted ultrasonic distress calls that stimulated the mother to behave more as she would have done in the natural environment. Rats handled in early development, or born to mothers that exhibit high levels of grooming, subsequently have lower levels of adrenocorticotropic hormone (ACTH) than the non-handled rats, indicating a major influence of the early experience on the hypothalamic–pituitary–adrenal (HPA) axis as well as on behaviour. The handled rats in the early studies were probably more like those found in natural conditions than the non-handled rats were.

Subsequently, the role of the mother rat in stimulating development of her pups has been studied extensively by Michael Meaney and his collaborators (Meaney, 2001). A mother that licks her pups a lot has offspring which, when adult, lick their offspring a lot. Conversely mothers who are low groomers have offspring who grow up to be low groomers. In this way, a characteristic style of maternal behaviour is transmitted from one generation to the next. Cross-fostering a pup born to a low-grooming mother to a naturally high-grooming mother switched the adult pattern of the pup to that of the foster mother (Champagne et al., 2003), showing that this is not a genetically transmitted trait but an acquired one. Liu et al. (1997) showed that the less a mother licked her pups, the higher the level of stress hormone release. Meaney's group showed that low licking and grooming is associated with a reduced number of glucocorticoid receptors (GR; the receptor for the active glucocorticoid, which in the rat is corticosterone and in humans is cortisol) in the hippocampus, thus reducing the efficiency of negative feedback from ACTH-stimulated corticosterone production on central neuroendocrine control of ACTH release. The epigenetic basis of this change is described in the next chapter. The point of this example is that the mother's behaviour towards her unweaned young can induce a behaviour in the offspring that is likely to be appropriate for them if the environment remains threatening when they are adults.

MATERNAL NUTRITION

When pregnant mother rats are given restricted diets, their offspring are smaller when they are born, but if these offspring are then given plentiful food they become much more obese than the offspring of mothers given an unrestricted diet (Jones and Friedman, 1982). This early observation has been followed by further extensive work on rats in many laboratories. Offspring born to undernourished rats develop increased appetites (Vickers et al., 2000), and show accelerated sexual maturation (Sloboda et al., 2009). Even though the undernourished rats are more sedentary when kept in standard laboratory cages (Vickers et al., 2003), their behaviour differs in another striking way from the control animals. When given a choice between pressing a lever to obtain food and running in a wheel, they are significantly more likely to run in the wheel (Miles et al., 2009). Whatever the explanation, the behavioural differences between rats that were undernourished during fetal life and those that were well nourished are remarkable.

The work on rodents is now being related to striking discoveries that have been made about human biology. Epidemiologists found that the smaller a baby at birth, the higher its risk of diabetes mellitus type 2 and cardiovascular disease later in life (Barker, 1998). An extensive field of enquiry then emerged, focusing on the relevance of early developmental plasticity in humans to the individual's subsequent health and risk of disease. This domain of research is often termed 'developmental origins of health and disease' (DOHaD) (Gluckman and Hanson, 2006). Those individuals whose mothers had been on a lower plane of nutrition had a greater propensity in later life to lay down fat and to develop insulin resistance and high blood pressure in an affluent environment; consequently, vascular and metabolic pathology was more likely to develop. Subsequent studies showed that such effects were not limited to those of small birth size and, indeed, birth size is now considered as simply a crude index of the fetal experience (Gluckman and Hanson, 2010). Indeed maternal experiences such as nutritional intakes can affect the offspring's biology independent of birth weight effects. Alterations in infant feeding also have effects later in life (Lucas and Sampson, 2006; Plagemann et al., 2009). We return to the functional and evolutionary significance of these observations in Chapters 6 and 7.

BEHAVIOURAL AND NEURAL PLASTICITY

Some learning processes, such as behavioural imprinting in birds and mammals and song learning in many species of birds, are restricted to sensitive periods early in life. We shall return to the ways in which robust mechanisms, generating the sensitive periods, interact with these forms of behavioural plasticity in the next chapter. The nervous systems of complex animals are readily changed by experience. This plasticity may involve growth or loss of connections between neurons, and malleability occurs throughout life. As with behavioural imprinting, some forms of neural plasticity occur early in development.

A striking example of neural plasticity has been demonstrated in the zebra finch (*Taeniopygia guttata*) which acquires its songs in the juvenile period. Roberts et al. (2010) were able to observe rapid changes in synaptic connections that occurred in an area of the brain associated with song learning. Using two-photon *in vivo* imaging, the dynamics of dendritic spines were measured immediately before and after finches were first exposed to tutor song in the juvenile period. Higher levels of spine turnover before tutoring correlated with a greater capacity for subsequent song imitation. In juveniles with high levels of spine turnover, hearing a song led to the stabilisation, accumulation and enlargement of dendritic spines within 24 hours of the experience. Moreover, *in vivo* intracellular recordings made immediately before and after the first day of exposure to song revealed robust enhancement of synaptic activity in the same brain region.

The lack of use by one sensory modality may involve the takeover of part of the brain by another modality. In individuals who have been blind from an early age, tactile cues stimulate parts of the primary visual cortex. The change in brain organisation was revealed by an experiment in which strong magnetic fields were used to disrupt the function of different cortical areas in people who had been blind from an early age. Disruption of the visual cortex disrupted their ability to read Braille or embossed letters. In contrast, transient disruption of the visual cortex in sighted people had no effect on their ability to perform tactile tasks (Cohen et al., 1997). Evidently the brains of the blind people had been reorganised in response to their particular experience of the outside world.

Clearly the developing organism has a particular capacity to demonstrate plasticity, and for many systems the capacity to be plastic later in life is much more limited both by functional considerations and because it may be energetically inefficient to maintain plasticity into later life. Nevertheless, some aspects of the phenotype, such as muscle or fat volume, are plastic throughout life. The most common form of plasticity in adults is seen in their behaviour. William Homan Thorpe (1956) brought together the insights of European ethology and holistic psychology with the vast corpus of work on the various mechanisms of learning from American and Russian laboratories, as well as those from psychology departments worldwide. Thorpe classified learning into five

categories: habituation, classical conditioning, instrumental conditioning, latent learning and insight learning. Some forms of learning such
as behavioural imprinting, which Thorpe discussed in his chapter on
insight learning, and the acquisition of song in birds may be restricted
to early development, but most can take place throughout life.

One of the most primitive changes in behaviour in response to
experience is non-specific. Sensitisation usually results from exposure
to an alarming stimulus (such as a blow-up toy snake suddenly becoming inflated), which elicits a variety of defensive or aversive reactions
from the animal. Subsequently, many other potentially aversive stimuli
(such as loud sounds) will have the same effect even though this would
not have been the case had the animal not been previously sensitised.

Habituation is defined as a decrease in response occurring as the
result of prolonged stimulation, which cannot be attributed to fatigue
or sensory adaptation. The phenomenon has been described widely,
from single-celled organisms to humans. In some cases, the underlying
process is simple and in other cases experiments suggest that the
subject establishes a specific representation of the stimulus in its nervous system. So if the animal has been habituated to a sound of a
specific duration and then tested with a sound that is shorter in duration, a non-habituated response is elicited when the end of the shorter
sound is reached. When a sound of longer duration is used, a non-
habituated response is elicited at the end of the habituated sound
(Sokolov, 1963).

Establishing a neural representation is key to the form of learning
that leads to a categorisation of the sensory world. Here again such
perceptual learning is found widely across the animal kingdom. In
humans, it leads to the recognition of faces and places. The ability to
distinguish between the vast array of objects, people and scenes experienced in a lifetime is of inestimable value and happens simply as a
result of exposure. The memory's capacity is extraordinary. In one
experiment, human subjects were shown up to 10,000 different slides
projected on to a screen in quick succession. Days later they were able to
recognise thousands of those images. Some of the images, such as a dog
smoking a pipe or a crashed aeroplane, were particularly striking. In
these cases the recall was even better and the subjects seemed to have a
virtually limitless capacity to store them (Standing, 1973).

The best-known type of associative learning process was made
famous by Ivan Petrovich Pavlov a century ago. Pavlovian (or classical)
conditioning allows the individual to predict what will be of real significance in the confusing world of sights, sounds and smells. Pavlov's

famous experiment was to teach a dog to expect food by repeatedly alerting it with a buzzer before the food was presented. Pavlov measured how much saliva the dog produced. As the dog was conditioned by the predictable association between the buzzer and the food, it came to produce saliva in response to the sound of the buzzer alone.

A different form of associative learning acts to control the environment. If a rat presses a bar and its action is swiftly followed by the delivery of food, it will repeat the action and will do so with increasing frequency if each time the action elicits the reward. As the action is strengthened, it may be repeated many times even in the absence of the reward (or 'reinforcer'). As learning proceeds, the conditions in which the action generates a reward become in themselves rewarding. Subsequently, the individual may perform a quite different act to achieve the conditions that are associated with getting the primary reward. By degrees a whole chain of different behaviour patterns can be established – a fact that is made use of by circus trainers (Coon, 2006).

In an idiosyncratic but highly original review, Bruce Moore (2004) identified 97 different processes of learning, some of which are highly complex, including abstract concept formation and cross-modal imitation. He may well have exaggerated but his survey serves to emphasise the heterogeneity of all those processes which are lumped under the general heading of 'learning'.

At one time psychologists were encouraged to interpret the behaviour of animals in the simplest possible way until they had good reason to consider otherwise (a precept also known as Lloyd Morgan's canon). In recent years, many people – for a variety of reasons – have rejected this advice, focusing on those aspects of animals that resemble the conscious behaviour of humans. This might seem questionable, given that much of human behaviour is unconscious and that humans differ from other animals in some respects, such as size and complexity of the nervous system. Nevertheless, the so-called cognitive revolution in the study of behaviour has undoubtedly greatly enriched the toolbox for understanding learning processes. We can do little more than hint at the extent of the discoveries here. What Thorpe (1956) called 'insight learning' related to, among other things, the experiments of Wolfgang Köhler (1925) on chimpanzees in the early part of the twentieth century. Among many other experiments, Köhler provided the animals with interlocking sticks. The chimps learned to assemble longer sticks which then enabled them to reach bananas suspended high above them and otherwise beyond their reach.

Modern examples include food caching by blue jays. If they have been observed while caching food by other birds they move the food, but not otherwise (De Kort et al., 2006). Rooks are able to reach initially inaccessible food floating on the top of water in a plastic tube by dropping stones into the water and thereby raising its level in the tube (Bird and Emery, 2009). Parrots have been found to develop abstract concepts of colour, shape or material content (Pepperberg, 2008). Ways in which one individual learns from another range from the simple examples of social facilitation of behaviour already established as a capacity in the individual, and enhancement of local cues, to cases where actions are observed and then copied (Laland and Galef, 2009). The study of complex cognition in animals, involving the plastic processes of learning, has expanded greatly in recent years and is very well summarised by Shettleworth (2010).

The most complex forms of learning clearly involve plasticity of considerable elaboration and it is not surprising, therefore, that a growing body of evidence links these capacities to the relative brain size of the species concerned (Reader and Laland, 2002).

IMMUNOLOGICAL PLASTICITY

In the immune system of humans and vertebrate animals, molecular plasticity takes the form of generating new antibodies to foreign proteins that hitherto have not been encountered by the individual. Antibodies are immunoglobulins used by the immune system to identify and neutralise foreign pathogens such as bacteria and viruses, preventing them from causing disease. A small region at the tip of the critical immunoglobulin is extremely variable in amino acid sequence, permitting the existence of millions of antibodies with slightly different tip structures. Each of these variants can bind to a different protein (or antigen). The diversity of antibodies allows the immune system to recognise an equally wide array of antigens (Male et al., 2006). Recognition and binding of an antigen by an antibody provides a marker that enables it to be attacked by other cells of the immune system.

The hugely diverse population of antibodies is generated by random combinations of a set of gene segments that encode different antigen binding sites, followed by random mutations in this area of the antibody gene, which create yet further diversity. Each form of antibody is made by a different clone of B-lymphocytes. The antibody is specific to the part of the antigen recognised by the immune system. Once a particular clone is stimulated by contact with an antigen, it undergoes massive proliferation to produce more antibodies and long-lasting immunity.

On the face of it, the foreign protein might have instructed the mechanism that generated the antibody in the sense that unique information carried by the protein provides the template on which the antibody is synthesised. However, the plasticity of the immune system involves selection rather than instruction since, by extremely rapid mutation and recombination within the histocompatibility complex, the immune system finds a match for the foreign antigen and this then sets in train rapid synthesis of the antibody from the mutated gene that provided the match (Neuberger, 2008). The gene has been selected by the challenge from outside the host's body.

THE NATURE OF PLASTICITY

The phenomena we have discussed in this chapter grouped under the general heading of 'plasticity' are without question biologically important, but do they have anything in common? As later chapters will detail, the central elements underlying many forms of plasticity are epigenetic processes, and plasticity operating at different levels of organisation often represents different descriptions of the same process. Underlying behavioural plasticity is neural plasticity, and underlying that is the molecular plasticity involving epigenetic mechanisms.

Edelman (1987) was attracted by the thought that the immune system provides a model for understanding the mechanisms that underlie learning. The plasticity of the immune system relies on a selective process. In contrast, at the level of the whole organism, the processes of learning that change behaviour seem to involve instruction. Whether the same selective process could be involved in any or all of the myriad examples of learning is, however, much more controversial. In the case of associative learning, for example, a cue from the environment instructs the individual about the causal nature of its environment, providing a link between something that is biologically significant and something that had hitherto been neutral. As yet, the underlying mechanism involving changes in neural connectivity is not readily attributable to a process that involves selection.

The image of selection might fit better with the examples of polyphenisms and reaction norms, whereby the genome is capable of giving rise to a variety of phenotypes depending on the individual's experience. In many cases of developmental plasticity that abound across the animal and plant kingdoms, the individual starts its life with the capacity to develop in a number of distinctly different ways. Like a jukebox, the individual has the potential to play a number of developmental tunes. The particular developmental tune played by the individual is triggered by a feature of the environment in which it has grown up – whether it is the odour of its predators, the available quality of food, or the presence of other males. Furthermore, the particular tune emanating from the developmental jukebox is adapted to the conditions in which it is played (Bateson, 1987; Bateson and Martin, 1999). The jukebox analogy has its drawbacks because it implies that the tune is preformed somehow (Oyama et al., 2001); but, like everything else, the predispositions have to develop. However, the image does draw attention to what may be a useful distinction between forms of plasticity that involve selection and those involving instruction.

Finally, it may be conceptually and mechanistically useful to distinguish between plastic responses to a challenge never previously encountered in the history of the species, and those involving previously encountered environmental challenges. The former may well present challenges for which the organism can have no evolved response, even though the induced accommodation enables the individual to survive and potentially to reproduce. The latter type of challenge, however, particularly if repeatedly encountered over evolutionary time, is likely to have led to functional adaptations. This may involve either immediate or predictive responses and may lead to the evolution of adaptive reaction norms or the capacity for developing polyphenisms. Such adaptations involve a suite of responses that are normally well suited to the conditions in which they are triggered. Nevertheless, the flexibility of biological organisation is such that one process that is well adapted to a given function might well be co-opted to serve a novel challenge.

CONCLUSIONS

This chapter has shown that plasticity can be viewed in many ways and along many different dimensions. Epigenetic mechanisms seem likely to underlie many of these processes and are discussed in later chapters. The temporal dimension, the dimension of different organisational levels, the mechanistic dimension of whether plasticity involves selection or instruction, and the functional and evolutionary dimensions (discussed in Chapters 6 and 7) are all part of the picture.

A multidimensional view is essential if the ways in which the organism responds to environmental cues and challenges are to be understood. However, the understanding needs to be broadened to take account of the ways in which plasticity is constrained and regulated. We address this concern in the next chapter.

5

Integration of robustness and plasticity

Up to this point we have dealt separately with the phenomena that we have grouped under the general headings of robustness and plasticity. However, the central message of this book is that the developmental processes that give rise to an individual's phenotype are not polar opposites, nor are they independent of each other. Indeed, plasticity is often regulated by robust mechanisms and robustness is often generated by plastic mechanisms. In this chapter we describe some aspects of how these processes are integrated.

The chapter opens by clarifying concepts of how the organism interacts with its environment. It then moves on to an account of sensitive periods in development, in which experience can have a particular impact on the phenotypic characteristics of an organism. The environmental cues are often highly specific; in the behavioural literature the animal's preferences are generally referred to as predispositions. Learning is the archetype of plasticity. Yet learning processes are highly regulated by robust mechanisms. These are discussed and then the chapter turns to the immune system, the workings of which have sometimes been likened to learning. Here again they involve the integration of robust mechanisms and adaptive plasticity. Finally, the molecular mechanisms of epigenetics are discussed in some detail since they are central to understanding how developmental mechanisms are integrated. At the end we return to our critique of the hard and fast distinction sometimes made between robustness and plasticity.

GENES AND THE ENVIRONMENT

One of the public health triumphs of genetics was the discovery that the expression of an inherited disease, phenylketonuria, could be prevented by providing the patient with an appropriate diet from an early age.

The affected people had a mutated gene that meant that an enzyme, phenylalanine hydroxylase, was deficient; therefore the amino acid phenylalanine could not be converted to tyrosine and instead was converted to toxic phenylpyruvate. The poison caused brain damage, resulting in great loss of cognitive ability, seizures, and other behavioural disorders. These dire consequences could be prevented by genetically testing neonates at birth and feeding children with the mutation a special diet that did not contain phenylalanine (Centerwall and Centerwall, 2000).

Another example of how the phenotypic consequences of a genetic change can be prevented by a change in the environment involved a knocked-out gene that normally codes for an important neurotransmitter (Rampon et al., 2000). When mice with the missing gene were kept in standard bare laboratory cages they developed a profound loss of the ability to recognise objects, loss of olfactory discrimination, and absence of memories of noxious events. However, when the mice were exposed daily to cages containing toys, running wheels and tunnels, and other forms of environmental enrichment, all these behavioural deficits were lost. The behavioural effects of the enrichment were accompanied by an increase in the synaptic density in a particular region of the hippocampus.

These examples of the interplay between genetic background and experience of the environment might seem to justify the phrase 'gene–environment interaction'. However, the developmental psychobiologist Danny Lehrman (1970) emphasised that it is the organism that interacts with its environment, not its genes. His message was not heeded, and many modern writers have persisted in referring to gene–environment interactions (G × E). We think the use of this phrase and its acronym are misleading because they are not inclusive enough, especially when they refer to what happens in individuals, as opposed to populations. An organism can change its environment without any obvious changes taking place in its genes. This has been termed 'niche construction' (Odling-Smee et al., 2003). Furthermore it can choose to inhabit environments that match its characteristics (Dover, 1986). The psychologists Scarr and McCartney (1983) referred to such behaviour as 'niche picking'. For these reasons we shall follow Lehrman and refer to organism–environment interactions when considering the integration of plasticity and processes that generate robust outcomes.

SENSITIVE PERIODS

In the 1920s, Charles Stockard, a leading embryologist of his time, observed that the embryos of many species could resist oxygen deprivation at some stages of their development, but that at times of rapid embryonic growth, oxygen deprivation could induce gross malformations such as two heads (Stockard, 1921). Tissues are particularly vulnerable to disruption at times of rapid growth. In humans much organ formation occurs during the first eight weeks after conception. The science of teratology is replete with examples of how toxins exert their effect at a specific stage of development; the impact of the drug thalidomide on limb development being one of the most famous. Another is the impact of maternal progestins (hormones related to progesterone) which were used in an attempt to prevent recurrent miscarriage. These progestins interfered with fetal androgen action and thus led to feminisation of the male genitalia. The degree to which the urinary opening was short of the tip of the penis depended on the stage of exposure to the progestin (Schwarz and Yaffe, 1980). Such examples of periods of vulnerability early in development to evolutionarily novel exposures often lead to phenotypic accommodation (see Chapter 4).

By way of contrast to the examples above, many sensitive periods in development exist within which the organism acquires information from its environment in order to develop normally or in a way that is appropriate to that environment. This information can play a crucial role in determining which of several alternative developmental courses the individual will adopt. For example, on the basis of the nutrition they receive in early life, female social insects are either capable of reproduction or adopt one or more sterile forms that perform specialised jobs in the service of the colony (Wilson, 1971; Figure 5.1).

The female honeybee larva may develop into either a queen or a worker morph, depending on what it is fed at an early larval instar stage; queens develop only if the larva is fed exclusively on royal jelly, a process known to involve epigenetic mechanisms (Kucharski et al., 2008). The genotype of the worker and queen that develop from the larva is the same but the two morphs differ enormously in subsequent phenotype due to this one environmental exposure within a sensitive period in development (Maleszka, 2008).

Defeminisation of the male rat brain occurs during the early postnatal period, up to about day 10 (Wright et al., 2010). The infant testis influences hypothalamic organisation through the secretion of testosterone, which is converted into oestradiol that in turn binds to its receptor in the brain and triggers a biochemical cascade. The consequence is that the adult male rat has patterns of gonadotrophic hormone secretion that are distinct from those of the female, and

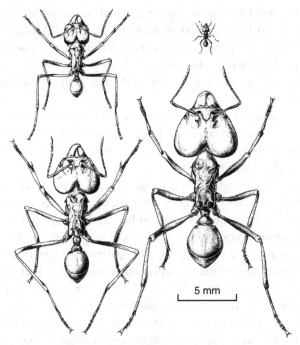

Figure 5.1. Sterile worker ants of the leaf-cutting species *Atta laevigata*, which has one of the most complex caste systems among the social insects. The ants are sisters from a single colony and each one has a specialised task to perform in the colony. The behavioural and morphological differences between them were induced by nutritional cues early in development. From Hölldobler and Wilson (1990), with permission from Turid Hölldobler-Forsyth.

subsequently exhibits mounting behaviour. Experimentally injecting testosterone into female rat pups can induce male development, but only within a narrow sensitive period between 7 and 12 days of age; no lasting effect comes from exposure to testosterone outside this period.

Unlike other vertebrates, sex is not determined by chromosomes in some reptiles, such as the turtles and crocodiles. Instead the temperature of the sand in which the egg is buried is crucial. Each individual starts life with the capacity to become either a male or a female. If the egg from which it hatches is buried in sand at a temperature below 30°C, the young greenback turtle becomes a male. If, however, the egg is incubated at above 30°C, it becomes a female. Temperatures below 30°C activate genes responsible for the production of male sex hormones and male sex hormone receptors; incubation temperatures above 30°C

activate a different set of genes, producing female hormones and receptors instead. It so happens that in alligators the sex determination
works the other way round, such that eggs incubated at higher temperatures produce males. The outcome depends on environmental
temperature during the middle third of embryonic development
(Yntema and Mrosovsky, 1982).

The developmental processes involved in the start of a sensitive
period correspond with changes in the ecology of the developing individual. These changes are linked to developmental processes of regulation and cellular replication. The processes that bring the sensitive
period to an end may reflect the passage of normal growth and temporal constraints on development in other related processes. Hypothalamic maturation in mammals needs to be largely completed before
weaning because the hypothalamus controls so many aspects of homeostasis, and maintaining plasticity in some pathways might constrain
fixation of others. Sometimes the terminating processes are related to
the gathering of crucial information and, except in extreme circumstances, do not shut down until that information has been gathered. In
these cases the ending of the sensitive period reflects the variable
opportunities for gathering such domain-specific information in the
real world. For example, in cold weather, ducks brood their hatched
young for longer than in warm weather, and the ducklings delay the
process of learning the characteristics of their mother (Bateson and
Martin, 1999). A limit must be set on such flexibility, however, because
so much else has to be done in development. If the relevant information
remains unavailable for too long, the individual may eventually have to
'make the best of a bad job' and develop without acquiring that
information.

Sensitive periods may be limited because of what has happened
previously to the individual. For example, structures in the brain that
have been altered by earlier experience may pre-empt the formation of
new structures. The brain mechanisms involved in visual development
have been analysed in cats and monkeys. The capacity of a cat's eye to
activate neurons in the visual cortex depends on whether that eye had
received visual input during the first three months after birth (Freeman
et al., 1996). If one eye is visually deprived during this period it largely
loses its capacity to excite cortical neurons. The other eye then becomes
dominant and, once established, usually remains dominant for the rest
of the individual's life. Once one set of neurons has established a
connection, they exclude others from doing so thereafter. Specific
neurobiological mechanisms, involving noradrenergic receptors at the

site of this plasticity, are required for such pre-emptive developmental changes to occur. Once a stable pattern of responding has developed normally, these mechanisms fall away. If the brain is not stimulated in the normal way through the visual pathways, then the reduction in the number of these receptors is delayed until the necessary interconnections between neurons have been established (Liu et al., 1994).

In general, neural plasticity during sensitive periods is likely to involve a variety of different mechanisms (Hensch, 2004). Moreover, as we have already noted, many such sensitive periods are not infinitely flexible. At some point, other developmental processes within the brain constrain plasticity. If the lack of visual experience goes on long enough, the period of sensitivity will terminate and the animal will end up with a brain that differs from the normal. In humans the development of binocular vision is seriously affected if a child has an uncorrected squint in the first few years of life. The child accommodates to the squint by becoming dependent on one eye and not using the other eye. The squint can be corrected by surgery, in which case the child usually develops normal binocular vision, but only if the surgery is performed sufficiently early in development. If it is left until much after the age of three years, surgery will do no good and the child will be left with permanently impaired vision in the affected eye (Campos, 1995).

The nature of such sensitive periods shows that experience at a given stage in development can fundamentally shape the individual's subsequent development. In general, the organisation that underlies the existence of such periods, such as in temperature-dependent sexual differentiation, is highly robust. In turn, after the plastic changes have been initiated during a sensitive period, the phenotypic outcome can be extremely robust, again as in the case of sex determination. The intertwining of developmental processes is complete.

PREDISPOSITIONS

Behavioural imprinting is the process by which a young animal rapidly learns the details of its mother's individual appearance and forms a social attachment to her (Bolhuis, 1991). Some young animals, such as goslings and cygnets, learn to recognise their father as well, but this is less common since, particularly in mammals, fathers rarely play a substantial role in caring for their offspring. A distinction is drawn between filial imprinting and sexual imprinting whereby the animal's experience later in life affects its sexual preferences. These sexual preferences are for partners that are slightly different from those individuals (usually close kin in natural conditions) with which the animal is already familiar (Bateson, 1983). Behavioural imprinting of both kinds provides a good example of a process that starts within a sensitive period. It also demonstrates nicely how learning is guided by certain robust predispositions.

The young animal is predisposed to set the learning process in motion, actively searching for objects with particular features in terms of colour, movement and shape. Stimulation in other modalities, when presented concurrently with visual stimuli, can have a powerful motivating effect. In domestic chicks and mallard ducklings, the sounds most effective in eliciting pursuit of a moving visual stimulus are maternal calls of their own species (Gottlieb, 1971). Auditory signals are important in guiding the process of forming a social attachment under natural conditions. Nonetheless, studies of filial imprinting provide unambiguous evidence for the formation of visual stimulus representations. Initially, domestic chicks and domestic ducklings have relatively unstructured social preferences at hatching. Movement was regarded as essential in 'releasing' the following response, and hence in initiating the imprinting process. However, the effectiveness of the many visual stimuli used in the imprinting situation depends on such properties as their size and shape, as well as on the angle they subtend, and the intensity and wavelength of light they reflect. Moreover, the rates at which these variables change are also important; hence the undoubted effectiveness of movement and flicker (Bolhuis, 1991).

Birds respond to a pattern of stimulation, and characterisation of the most effective stimulus must be cast in terms of clusters of features. Some features of the jungle fowl, the ancestral form of the domestic fowl, are particularly attractive to chicks. The head and neck is the crucial characteristic, but the head and neck of a small mammal is as effective as that of a jungle fowl (Horn and McCabe, 1984). In Japanese quail, the movement of a live adult female has a powerful motivating effect on the response to her by the chicks (ten Cate, 1986). Newly hatched domestic chicks, at their first exposure to animation sequences made up from points of light, exhibit a spontaneous preference for biological motion patterns over non-biological patterns (Vallortigara et al., 2005). All these examples show how plasticity is initiated by robustly developing predispositions, and that once strong attachments are formed, they may remain stable for the rest of the animal's life.

Another striking example of the ways in which predispositions influence learning is in the acquisition of songs by birds. The process starts early in life. The typical pattern is for the young male bird to listen to and store sounds made by his father and other males during the first few months after he has hatched (Marler and Slabbekoorn, 2004). The following spring he produces a range of sounds and, by degrees, settles on songs he has heard before. When he is mature, he uses his songs to defend his territory and attract females. The range of songs acquired by each male is transmitted within the neighbourhood from one generation to the next, in much the same way as language, customs and ideas are transmitted across the generations in humans. Under laboratory conditions, hand-reared birds will learn sounds played back from tape-recorders. Typically, however, they are much more likely to acquire songs made by their own species than songs made by other species, even when the songs are mixed up in the recordings (Marler and Peters, 1977). Marler (2004) referred to the readiness of birds to learn about particular features of their acoustic environment when developing their songs as an 'instinct to learn'. Despite the unfortunate ambiguity of the term 'instinct' (see Chapter 2), the concept of a hunger to learn about certain things is important when considering the interplay between plasticity and robust features of the animal.

THE STRUCTURE OF LEARNING

Learning is the most obvious way in which individuals after birth interact with, and are changed by, their environment. Learning is entwined in the processes of behavioural development, adapting individuals' behaviour to local conditions, enabling them to copy the behaviour of more experienced individuals, and fine-tuning preferences and actions that were inherited from previous generations.

As any dog-owner knows, a hungry dog will do many other things once it detects cues that predict the arrival of food: it will go to the food bowl, whine, wag its tail, jump up, and show all the familiar signs of expectation. When learning about the relevant cue that predicts the arrival of food, the sequence in which the events occur is crucial. If the cue comes *after* the presentation of food to a dog that has not yet been conditioned, it will not salivate or show any other expectant signs when the cue is subsequently presented. The link in time between the action and the outcome is crucial.

The interplay between the rules for learning and robust features of the animal's behaviour can sometimes lead to amusing outcomes, as was described by Keller and Marian Breland (Breland and Breland, 1966). They were experimental psychologists who later became professional animal trainers. On one occasion they tried to train pigs to take a wooden coin from a pile and drop it into a piggy bank in return for a food reward. Initially the pigs learned the task well, but as they came to associate the pile of coins with food, the sequence broke down because the pigs started rooting in the pile of coins!

Storing information about visual experiences requires a different set of rules. Animate and inanimate objects in the real world are rarely flat and their appearance depends on how they are viewed. A friend or relative is easily recognised from the front or the back, whether they are in the distance or close up. But they may not be so readily recognised if the photograph is taken from an odd angle such as from their feet. The recognisable features of a familiar person are fused together by the brain into a single category when these different views are seen in quick succession (Bateson, 2000). Time plays a different role in such perceptual learning than is usually the case in Pavlovian conditioning. The order in which different events are experienced is important when one event causes the other, but unimportant when the experiences are of different views of the same object. In the case of perceptual learning, just as in conditioning, profoundly important plastic processes are dependent on robust rules.

IMMUNE SYSTEMS

Immune systems represent another example where their effectiveness relies on plasticity but in turn is dependent on the robustness of cellular organisation (Tieri et al., 2007). The immune systems of vertebrates consist of many types of proteins, cells, organs and tissues that interact dynamically and change over time. The so-called 'adaptive immune system' becomes able to recognise previously encountered pathogens and antigens, as described in the previous chapter. Immunological memory created from the initial response to a pathogen (or more generally an antigen) leads to antibody formation, which enables the organism to fight that particular pathogen on subsequent encounters. The cellular and humoral immune systems are then able to distinguish between the cell surfaces of foreign invaders and those of the body they serve – in other words, between self and non-self. The more accurately the immune system performs both these functions, the better the individual is at avoiding infection and disease. If, on the other hand, the immune system is not robust in recognising self, it may attack tissues in its own body, causing autoimmune diseases such as rheumatoid arthritis.

THE RISE OF EPIGENETICS

Modern understanding of an individual's development goes well beyond accepting that interactions between the organism and its environment are crucial. The conditional character of an individual's development emphasises the need to understand the processes of development that underlie these interactions. This is what Waddington (1957) called the study of epigenetics. More recently epigenetics has become mechanistically defined as the molecular processes by which traits defined by a given profile of gene expression can persist across mitotic cell division, but which do not involve changes in the nucleotide sequence of the DNA. The term has come to describe those molecular mechanisms through which both dynamic and stable changes in gene expression are achieved, and ultimately how variations in environmental experiences can modify this regulation of DNA (Jablonka and Lamb, 2005; Gilbert and Epel, 2009). It is one of the most rapidly expanding components of molecular biology. Robust mechanisms of development and those that generate plasticity are closely intertwined by such mechanisms.

Variation in the context-specific expression of genes, rather than in the sequence of genes, is critical in shaping individual differences in phenotype. This is not to say that differences in the sequences of particular genes between individuals do not contribute to phenotypic differences, but rather that individuals carrying identical genotypes can diverge in phenotype if they experience separate environmental experiences that differentially and permanently alter gene expression. We shall describe what is already known about the molecular mechanisms of epigenetics in some detail. This topic is at the forefront of modern developmental research and recurs throughout this book.

MOLECULAR EPIGENETICS

The level of expression that occurs at any particular gene is ultimately determined by the accessibility of the DNA to the RNA polymerase II enzyme and those other factors that facilitate the transcription of the gene, giving rise to messenger RNA. In turn, the ability of these enzymes and transcriptional modifiers to gain access to the DNA is dependent on the structure of the chromatin and, in particular, how tightly it is

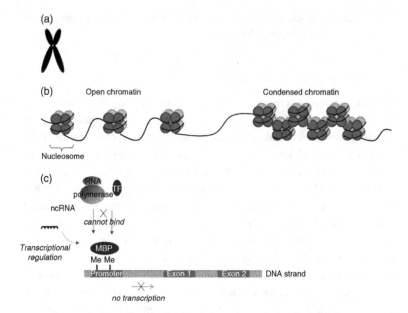

Figure 5.2. The chromosome and the epigenetic regulation of gene expression. (a) Schematic of a chromosome, which consists of densely packed chromatin. (b) Chromatin is made up of nucleosomes, which are complexes of DNA wrapped around a group of eight histone proteins. Various reversible chemical modifications to histones can affect how tightly the DNA is associated with the proteins. This, in turn, determines the access of transcriptional machinery and regulatory RNAs to the DNA within: open chromatin is amenable to gene expression, while condensed chromatin is repressed. Histone acetylation is generally associated with transcriptional activation, while other types of modifications, such as methylation, result in context-dependent effects – that is, gene activation or repression may occur depending on the amino acid residue being modified. (c) Epigenetic regulation of a gene may also occur by DNA methylation (Me). In general, methylation at the promoter region, as depicted in this figure, results in the recruitment of methyl-binding proteins (MBPs) and exclusion of RNA polymerase and transcription factors (TF), thus inhibiting transcription. Removal of the methylation mark enables transcription to resume. In contrast, methylation at some intragenic regions has been associated with increased gene expression. Non-coding RNAs (ncRNA) have varying regulatory effects at the transcriptional and translational level.

packed. Chromatin consists of chromosomal DNA coiled around a protein core; the proteins involved are called histones and the nature of the wrapping is under complex biochemical control (Figure 5.2). Tightly wrapped or condensed chromatin means that the DNA grooves where transcriptional factors normally bind are not accessible to those factors. Open chromatin, in contrast, can bind regulatory proteins and RNAs (Mohd-Sarip and Verrijzer, 2004).

Epigenetic mechanisms may involve either enzymatic modification of the DNA or histones, or altered processing, editing and expression of small non-coding RNAs which can coat the chromatin. These overlapping and interdependent processes change the accessibility of transcriptional regulating proteins and complexes to the DNA, and so influence the transcription of DNA (Cedar and Bergman, 2009; Mondal et al., 2010).

One of the best-studied processes of epigenetic regulation is that of DNA methylation. This is triggered by the activation of the DNA methyltransferase enzymes (DNMTs). One class of DNMT induces the methylation change and a second class maintains these changes across mitosis. Generally the methylation occurs at DNA sequences involving the nucleotides cytosine and guanine (termed CpG sequences). These sequences may occur in clusters known as CpG islands or as more scattered sequences which may or may not be methylated. The generally accepted view is that methylation of a CpG site leads to gene repression by attracting binding proteins and inhibiting binding of transcriptional regulators to that region of DNA (Li and Bird, 2007). However, the impact on gene expression appears to be dependent on the site of methylation: while methylation at 5' promoters is generally, but not always, associated with repression, intragenic methylation leads to either repression or enhancement of gene expression (Jones, 1999; Lorincz et al., 2004), and this phenomenon appears to be more widespread than previously thought. The methylated cytosines attract methyl-binding proteins (MBPs) which further affect the conformation of the chromatin. Methylated cytosines can also be converted to hydroxymethylcytosine, particularly in the brain (Tahiliani et al., 2009). Histone modifications are the other major class of epigenetic change. Amino acid tails within the histone protein can be subject to enzymatic modification, thus changing the structure of the chromatin and, in turn, the accessibility of binding proteins and transcriptional regulators. Histone modifications interplay with DNA methylation in modulating chromatin structure. The nature and the functional significance of such modifications is highly specific (Jenuwein and Allis, 2001).

DNA sequences that do not code for proteins were previously called 'junk DNA' and comprise the bulk of the mammalian genome. However, much of this codes for RNAs that are not translated into proteins. The recognition of the critical biological significance of these sequences has been one of the major discoveries of recent years. Many non-coding RNA molecules act as regulatory factors either by association with intra-nuclear proteins or by binding to the DNA. They are likely to play a major role in conferring specificity to these epigenetic processes. For example RNA molecules contain cytosines that can be methylated or hydroxymethylated, perhaps under environmental influence. Small RNAs may be changed under enzymatic control, with cytosine converted to uracil or adenosine to inosine (Mattick, 2010), affecting the way in which they operate – a phenomenon known as RNA editing. The small non-coding RNAs bind to the DNA, affecting enzymatic access to the chromatin, and consequently provide specificity to epigenetic processes (Koziol and Rinn, 2010).

These molecular processes involved in phenotypic development were initially worked out for the regulation of cellular differentiation and proliferation. All cells within the body contain the same genetic sequence information, yet each lineage has undergone specialisations to become a skin cell, hair cell, heart cell, and so forth. These phenotypic differences are inherited from mother cells to daughter cells. The process of differentiation involves the expression of particular genes for each cell type in response to cues from neighbouring cells and the extracellular environment, and the suppression of others. Genes that have been silenced at an earlier stage remain silent after each cell division. Such gene silencing provides each cell lineage with its characteristic pattern of gene expression. Since these epigenetic marks are faithfully duplicated across mitosis, stable cell differentiation results. These mechanisms also have many other roles in development, including mediating many aspects of developmental plasticity (see Chapter 8).

MECHANISMS UNDERLYING INDUCTION OF PHENOTYPIC DIFFERENCES

In honeybees, while the reproductive queen and the sterile workers both have the female genotype, they have strikingly different phenotypes due to different suites of genes being expressed; the phenotypic difference is induced by nutrition of the larvae at a critical stage in development. As can be seen in Figure 5.3, if the DNMT enzyme necessary for *de novo* methylation of DNA is inhibited, a proportion of the larvae develop as queens, demonstrating the epigenetic basis for this nutritionally induced polyphenism (Kucharski et al., 2008).

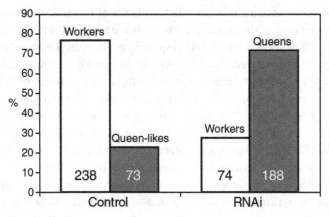

Figure 5.3. The morphological and behavioural differences between the reproductive queen honeybee and her sterile worker sisters are great. The differences are induced early in development by nutrition, and the epigenetic effects are striking. Different suites of genes are expressed. If production of the DNMT enzyme necessary for *de novo* methylation of DNA is inhibited by small interfering RNA (RNAi), a proportion of the larvae develop as queens. Numbers in bars refer to adults in each phenotypic category. From Kucharski et al. (2008), reprinted with permission from the American Association for the Advancement of Science (AAAS).

In vertebrates, specific patterns of gene expression within individual populations of neural cells can be dynamically and permanently altered through variations in early experiences during development. Much of this work has come from studies on laboratory rodents. Variations in mother–infant interactions during early life lead to long-term changes in behaviour; this arises due to variations in brain development induced by epigenetic modification of gene expression in the brain (Champagne and Curley, 2009; Champagne, 2010). The daily separation of rat dams from their offspring for prolonged periods of time results in these individuals exhibiting a much higher stress response as adults, which is partly mediated by high expression of the hormone vasopressin in the paraventricular nucleus of the hypothalamus (Murgatroyd et al., 2009). These neurons, in concert with corticotropin-releasing hormone (CRH) neurons that also reside in the same region of the brain, stimulate the release of adrenocorticotropic hormone (ACTH) from the anterior pituitary, which then stimulates the release of corticosterone from the adrenals. In mice, these long-lasting increases in vasopressin are due to reduced levels of DNA methylation of the relevant gene, specifically in this brain region (Murgatroyd et al., 2009).

Those rat dams that provide high levels of tactile stimulation to their offspring in the form of maternal licking have offspring that, as adults, show a decreased stress response and increased expression of the glucocorticoid receptor (GR) protein in the hippocampus (which facilitates a more efficient negative feedback of the stress response) (Weaver et al., 2004). These differences are not inherited genetically from their mothers. When rat offspring are cross-fostered, it is the maternal behaviour of the foster dam, not their biological dam, that predicts their future phenotype. Receiving high levels of licking neonatally is associated with reduced levels of DNA methylation of the promoter region of the GR gene, which is established by day 6 postnatally and persists throughout adulthood. Moreover, this change in DNA methylation is specific to those GR genes located within the hippocampus and not elsewhere in the brain, demonstrating that changes induced by early experience are relatively specific. Tactile stimulation that is received early in life stimulates various intracellular events in the hippocampus, including elevated levels of a particular transcription factor that activates the GR promoter and prevents it from being methylated. Offspring treated later in life with an agent that reversed the epigenetic modification resulting from low licking by their mothers switched to high levels of offspring licking themselves (Weaver et al., 2005).

We described in Chapter 4 how the mammalian mother's nutritional state affects the phenotypic characteristics of her offspring. Pregnant rat dams placed on low-protein diets have offspring with elevated GR gene expression in the liver, which is associated with decreased DNA methylation of the gene promoter and is partially responsible for the altered metabolic phenotype of these offspring (Lillycrop et al., 2007). The quality of the maternal diet can also lead to long-term changes in offspring gene expression. Mice that carry a particular copy of the *agouti* gene typically possess a yellow coat colour and obese phenotype. If the mothers of these mice receive dietary supplementation with methyl donors such as choline and folate during pregnancy, then the expression of this gene copy becomes epigenetically silenced via DNA methylation, and the mice develop a lean phenotype and a normal, brown coat colour (Waterland and Jirtle, 2003).

The role of epigenetic processes involved in some of these effects in humans is the subject of intense research. Individuals whose mothers were exposed around the time of conception to famine have reduced DNA methylation of the gene that codes for insulin-like growth factor 2 (IGF-2) compared with their siblings conceived before or after the famine (Heijmans et al., 2008). Furthermore, the degree of methylation of a CpG sequence in the promoter region of a gene involved in fat development is associated with maternal diet in early pregnancy and, in turn, is correlated with insulin sensitivity and body composition in later childhood (Godfrey et al., 2009). Increases in the DNA methylation of the promoter region of the GR gene have been reported in the blood cells of human infants that were born to mothers with depression during the third trimester of pregnancy, with these changes in GR gene methylation being correlated with cortisol levels between infants (Oberlander et al., 2008). Gene-specific methylation changes at birth have been associated with prenatal exposure to environmental toxins such as cigarette smoke (Guerrero-Preston et al., 2010) and traffic-related air pollutants (Perera et al., 2009). In the latter case, methylation of the ACSL3 gene was significantly associated with childhood asthma symptoms. Thus variations in the maternal environment during gestation can have long-term effects on development through alterations in gene expression that are sustained via DNA methylation.

We have been concerned in this chapter with the interplay between those processes that generate robust outcomes and those that respond plastically to changes in the environment in the individual. The issues raised by the studies of populations are entirely different (see Chapter 2). Nevertheless, epigenetic research does raise questions about the ways in which data obtained from populations are interpreted. For example, in both medical and psychological research, attempts have been made to separate genetic from non-genetic effects by the comparison of identical and non-identical twins. The correlation between a trait in 'identical' monozygotic (MZ) twins, who share all their genes in common, is compared with that in 'non-identical' dizygotic (DZ) twins, who do not. The difference in correlation between the MZ twins and that between the DZ twins is used to calculate the likely contribution of genes to that trait. The likelihood was, however, that MZ twins shared a placenta and therefore had a more similar uterine environment than did DZ twins. The epigenetic profiles of MZ twins are similar at birth, but as they get older the profiles progressively diverge (Fraga et al., 2005; Wong et al., 2010). This finding points to the similarity of the uterine environment of MZ twins and may explain some of the differences in the correlations found in the two groups of twins. Clearly, interpretation of twin studies is fraught with difficulties and the limitations of traditional G × E attempts to partition sources of variation in populations have already been discussed in Chapter 2. The epigenetic data do not contradict the finding that MZ twins become more similar in their cognitive ability as they grow older (Davis et al., 2009), although this finding is plausibly explained if, as they grow older, the MZ twins share a more similar cognitive environment than do the DZ twins.

CONCLUSIONS

Labelling the characters of an individual as developmentally robust or as developmentally plastic has been seen by some as a useful way of classifying the characteristics of an organism. According to such a view, the anatomical structure of the adult might be seen as robust and its behaviour as plastic. However, the evidence presented in this chapter shows how a range of alternative anatomical structures may be generated by conditions in early life and how plasticity is governed by robust rules such as those used for detecting causality in the environment. Development involves both internal regulation and reciprocity with the environment (Bateson, 1976). Careful analysis of what happens during development suggests that it is no longer helpful to retain a hard and fast distinction between robustness and plasticity.

The precise way in which a variety of developmental processes are integrated depends on the utility of the phenotype which they generate. In the next chapter we discuss how the outcomes will be advantageous in some ecological conditions and less so in others.

6

Current function of integrated developmental processes

The focus in this chapter is on explaining the current benefit to the individual of possessing those characteristics of integrated development that were discussed in the previous chapter. If an individual deviates significantly from the adapted norm or, because of changes in local circumstances, its characteristics no longer match the requirements of the environment, then a measurable loss of benefit should result. If no such loss occurs, then arguably that character has no current utility in terms of survival or reproductive success. Monaghan (2008) has carefully reviewed the types of experiments that distinguish between the competing hypotheses attempting to address these issues.

We discuss first the utility of maintaining the form and coherence of the whole body. We then turn to the utility of 'coping' mechanisms, and then to the special importance to survival and reproductive success of the plastic processes that operate early in life and that generate relatively robust phenotypes. The benefits may accrue long after cues from the environment or the mother have established a particular trajectory for development. Behavioural plasticity continues into adult life and, in the instances of the many forms of learning, the benefits to the individual can be clear-cut. Sometimes phenotypes established by plastic mechanisms are poorly matched to the environment in which the individual finds itself. These cases, which are important in understanding loss of health in humans, might seem like a challenge to the notion of utility. However, on the contrary, they provide a demonstration that the developed phenotype normally provides great benefit for the individual.

MAINTENANCE OF FORM

Maintaining the coherence of a body required to perform many different tasks is an astonishing feat of integrated development.

The adaptedness of the many characteristics of the organism to those jobs is crucial to its survival. In Chapter 3 we discussed the various mechanisms that exist to promote robust maintenance of form even when the environment is variable. Blumberg (2009) gives many instances of the ways in which developmental processes do occasionally deviate from the norm. The anecdotal literature on so-called 'monsters' extends way back into history and provides rich material. Charles Darwin's mentor, John Stevens Henslow, was keenly interested in 'monstrosities', and at the end of the nineteenth century, William Bateson (1894) provided extensive catalogues of examples. Some specific cases are discussed in Chapter 8, where we examine the impact of development on evolution, but the usually manifest disadvantages to individuals who have suffered developmental disruption emphasise how important normal development is to both health and reproductive success. Important exceptions to this rule have the potential to launch new lineages (Blumberg, 2009). Usually, however, disruption of development is a major cost to the organism, even if the disruption does not lead to death.

The problems of integrating the development of the vast range of capacities required by an organism are compounded when the organism is the product of the gametes of two independent parents. For sexually reproducing species, the compatibilities of the mate's genotype and phenotype are both prerequisites for successful mating. Success may depend on compatibility of behaviour or anatomy in the case of animals, but above all it depends on a match between the genome of the egg and that of the sperm if development is to proceed normally. Mating with an individual too different from oneself carries costs when the offspring are sterile, as when a horse mates with a donkey to produce a mule. A more subtle cost arises when parents of the same species have adaptations to different local conditions, or when their own survival depends on the coadaptation of specific genes and the combination of genes is different in each parent.

When the life history of the species demands careful nurturing of the offspring, the parents may go to a lot of trouble to mate with the best partner possible. A mate should be not too similar to oneself but not too dissimilar either. Mating with an individual too close in its characteristics to oneself loses the benefits of sexual reproduction, including creating new combinations of genes from different parents, which improves the effectiveness of the defence mechanisms against pathogens. Certainly, in many plants and in some animals asexual

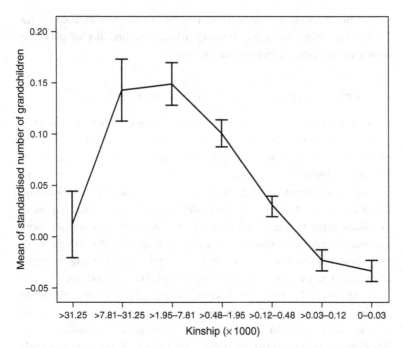

Figure 6.1. Relationship between the degree of kinship of partners and the number of their grandchildren in an Icelandic population. The greatest number of grandchildren was produced by partners who were third or fourth cousins, indicating that both inbreeding and outbreeding reduce reproductive success. Much evidence suggests that the choice of a mate is dependent on experience in early life, with individuals tending to choose partners who are a bit different but not too different from familiar individuals, who are usually close kin. A robust mechanism required for the integrated development of the offspring's phenotype depends on a plastic process in the parents. From Helgason et al. (2008), reprinted with permission from the American Association for the Advancement of Science (AAAS).

reproduction does occur, but costs are often incurred if this persists for several generations, and such organisms may be vulnerable to parasitism and changes in the environment. Animal studies have suggested that an optimal degree of relatedness is most beneficial to the organism in terms of reproductive success (Kalbe et al., 2009). A study of a human Icelandic population also points to the same conclusion. Couples who were third or fourth cousins had a larger number of grandchildren than more closely related or more distantly related partners (see Figure 6.1; Helgason et al., 2008).

This discussion brings us to the functional significance of the broad class of phenomena clustered under the heading of plasticity which we introduced in previous chapters.

COPING WITH DEVELOPMENTAL DISRUPTION

Development sometimes seriously deviates from the norm. Mutation of genes is an important source of such disruption (Leroi, 2003). For example, achondroplasia leading to gross skeletal disruption and dwarfism occurs from either spontaneous or inherited mutations in a gene coding for a receptor for a growth factor (Vajo et al., 2000). In addition factors external to the individual can seriously upset development. These are often toxins or infections to which the lineage has not been exposed or has not evolved with a coping response. The rubella virus may cross the human placenta and infect the fetus, leading to severe brain damage, cardiac malformation and growth retardation.

Thalidomide, taken by women in the middle of the twentieth century to relieve morning sickness during pregnancy, is metabolised and produces a toxin that interferes with limb formation of the fetus. Many other agents including anticancer drugs may similarly interfere with fetal development. Although the costs of such disruption can be enormous, depending on the abnormality induced, affected individuals may cope remarkably well. Having accommodated, they may well survive to reproductive age.

IMMEDIATE ADAPTIVE RESPONSES IN EARLY DEVELOPMENT

We suggest that, in terms of mechanism, a distinction should be drawn between accommodation in response to evolutionarily novel stimuli such as toxins that disrupt development, and the responses to severe environmental conditions that have occurred previously over evolutionary history and have led to the evolution of responses that are adaptations to those conditions (Bateson et al., 2004; Gluckman et al., 2005b). As well as being associated with disruption of the normal processes of development, the former do not involve specific adaptive responses to the environmental challenge. In contrast, the latter class involves mechanisms for balancing short-term survival and the long-term impact on reproductive success. Such trade-offs lie at the heart of life history theory. A good example, described in Chapter 4, is that of the tadpole of the spadefoot toad, which responds to signals indicating that the pond in which it is living is drying out (Denver et al., 1998). It undergoes early metamorphosis to a terrestrial form even though this results in a smaller mature body size that leaves it at greater risk of predation. Intrauterine growth retardation occurring in a mammalian fetus when transplacental nutrient supply is limiting is an appropriate adaptive response by the fetus, redistributing limited nutrients to protect the growth and function of its heart and brain at the expense of growth of the rest of its body. The fetus is more likely to survive to birth, but smaller neonates are less likely to survive and, when they do, have a reduced chance of breeding successfully.

ANTICIPATORY AND DELAYED RESPONSES

Other developmentally induced responses can lead to phenotypic changes that may either manifest later in the life course or their potentially advantageous significance appears to manifest at a later stage in the life course than when it was induced. In general, the responses of delayed potential benefit can lead to one of two classes of outcomes in later life: distinct morphs in polyphenic species, or those where the traits of interest show continuous variation, forming a norm of reaction. A key feature of both forms of plasticity is that while the relevant developmental decisions are of necessity made in early life because of constraints on how development can proceed beyond periods of sensitivity, the potential adaptive advantage of such plasticity is presumed to be obtained later in life. In other words the organism responds plastically in 'anticipation' of the likely conditions of its future environment (Bateson, 2001). One term used to describe such plasticity is that it employs 'predictive adaptive responses' (Gluckman et al., 2005a).

In forecasting the future environment, the organism responds in early development to cues in a characteristic manner so as to produce an integrated phenotype that is appropriate to the predicted environment. In Chapter 3 we described how the migratory locust exists in a migratory form – as its name implies – or, at the other extreme, in a solitary form. The two are anatomically, biochemically and behaviourally distinct – so distinct, indeed, that they were sometimes regarded as different species. The adaptive significance of the plasticity is that the solitary form is more appropriate when nutritional sources are rich, while the migratory form is more appropriate when nutritional sources are limited in the habitat in which its larval development occurred. The triggering cues are those that predict population density. The anticipatory nature of the response is illustrated by the early induction of wing shape (larger in the migratory form), which has no value until later in life when the mature form is reached (Applebaum and Heifetz, 1999).

Alternative forms within a species may develop in vertebrates. For example, salamanders may have either an aquatic or a terrestrial form depending on current and predicted ecological conditions. The coat thickness of the meadow vole (*Microtus pennsylvanicus*) is determined before birth, depending on the season in which it is born (Lee and Zucker, 1988), with those born in the autumn having thicker coats. This is an alternative developmental trajectory, where the advantage of a thicker or thinner coat is not apparent at the time of the developmental decision, but rather is seen some time later when the offspring have to contend with warm or cold weather. The cue that leads to adoption of a thin or thick coat trait is mediated through the mother, whose exposure to changing day length affects the synthesis of melatonin (Lee and Zucker, 1988).

In each of the cases, a developmental decision must be made in early life to proceed down the pathway to one form or another. For example, in the locust, the alternative form has different mouth parts and wing sizes, and the commitment to one form or the other must be made early so that these develop appropriately. This reflects the constraints on maintaining continuous plasticity, which limit the opportunities for anatomical and functional plasticity as the organism ages. The cue must be informative but not necessarily directly operating on the traits that change. For example, in the case of the meadow vole, while the inducing cue is day length, the response system relates to thermal homeostasis. The adaptive advantage of such predictions is obvious if the fidelity of the prediction of the future environment is high: they leave the organism better placed to survive and reproduce in the anticipated environment (Moran, 1992). These examples, primarily based on polyphenic morphs, provide a special case of the more general phenomenon where the outcome of development is not one of a series of discrete phenotypes but instead a continuous range of outcomes (i.e. a continuous reaction norm).

In the African butterfly *Bicyclus anynana*, food restriction at the larval stage improves the capacity of the adult butterfly to survive food restriction; thus, fitness is enhanced by predictive adaptation (Bauerfeind et al., 2009). Similarly, when the eggs of the ringed salamander (*Ambystoma annulatum*) are exposed to chemical cues of predators, the post-hatching larvae show greater shelter-seeking behaviour (Mathis et al., 2008).

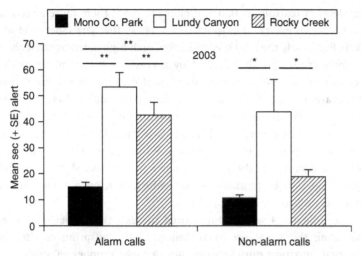

Figure 6.2. Belding's ground squirrels are less nervous when reared in open habitats (Mono Co. Park and Rocky Creek), where the risks of predation are lower, than when reared in woodland habitats (Lundy Canyon), where the risks are higher. Responses were measured by the time spent engaging in alert behaviour after playbacks of alarm and non-alarm calls. The differences may be induced by the behaviour of the mother, as is the case for laboratory rats. From Mateo (2007), reprinted with permission from Springer Science+Business Media.

In Chapter 4 we introduced the body of observations demonstrating that, in rats, mothers exhibiting low levels of maternal care have offspring that express a high level of fearfulness in response to stress. These offspring themselves grow up to be low-licking mothers and thus recreate the same patterns of behaviour in the next generation. These changes are associated with epigenetic changes in the expression of the glucocorticoid receptor in the brain and heightened activity in the hypothalamic–pituitary–adrenal axis. The system is adaptive if an animal that has been predicted a stressful environment by virtue of its mother's behaviour is more likely to survive if it is alert to the threat of predators.

Analogous responses may be seen in the wild. Belding's ground squirrels (*Urocitellus beldingi*) living in open grassland habitats develop into less fearful adults than those living in closed woodland habitats (see Figure 6.2). The dangers of predation are lower in the open than in woods, (Mateo, 2007). Striking differences are also found within arctic animals such as the lemming and snowshoe hare. In the latter, when the risk of predation is high the mother gives birth to offspring with a high level of fearfulness that is thought to make them more alert and better able to escape from predators such as the lynx (Sheriff et al., 2009). Eventually the prey species suffers such high mortality that their numbers crash, quickly followed by a crash in the numbers of the predator species. In the next generation of pregnant hares, the threat of predation is low. Some carry-over from the grandmothers' stressed state occurs (Sheriff et al., 2010) but eventually the offspring are predicted a safe environment and thus have a lower flight response when they are independent of their parents. When the predator population increases once again, the pregnant hares face increased stressors and signal that danger to their fetuses, which develop a higher flight response. And so the cycle continues.

In Chapter 4 we also introduced the extensive series of studies addressing the long-term consequences to the offspring of altering maternal nutrition during pregnancy. A large number of studies in the rat have shown that reducing maternal nutrition leads to offspring with a phenotype that favours the development of hyperphagia, fatty preference in the diet, insulin resistance and obesity (Vickers et al., 2000, 2003). The argument is that this represents a predictive adaptive response where the fetus anticipates an adverse postnatal nutritional environment and alters its physiological development accordingly, so as to be better placed to cope in a low-nutrition environment (Gluckman and Hanson, 2005). This hypothesis has been tested by changing the prediction at birth by injecting infant rats with the hormone leptin, which signals a high-nutrient environment (Vickers et al., 2005). The injected offspring do not develop the altered metabolic phenotype as adults, and the associated expression and epigenetic changes are also reversed (Figure 6.3; Gluckman et al., 2007c). Such data provide empirical support for the predictive hypothesis, since the developmental outcome is changed by experimentally manipulating the predicted environment.

The original epidemiological observations of David Barker and colleagues (Barker et al., 1989) focused on an association between birth weight and health in later life. However, low birth weight represents an extreme in which the fetus has also made an immediate adaptive response (that is, its birth weight has been reduced). Indeed, where the nutritional conditions for the mother are so poor that the fetus has to make adjustments to cope with these conditions, it would be inappropriate to describe such a response as a prediction of future conditions. The fetus is simply 'making the best of a bad job' through an immediate adaptation (Gluckman et al., 2005b) and may have to

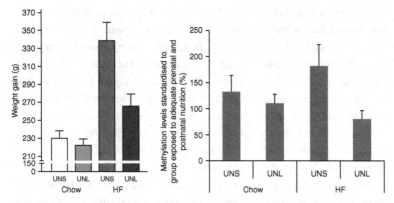

Figure 6.3. Comparison of weight gain (left) and methylation levels of the promoter of a transcription factor involved in metabolic regulation (right), in rats subjected to prenatal undernourishment (UN) then exposed to a postnatal chow or high-fat (HF) diet. Prenatally undernourished–postnatally overnourished rats present a classic example of mismatch. Neonatal treatment with leptin (L) reverses the weight gain and epigenetic changes seen in the mismatched rats, as compared to a sham treatment with saline (S). From Vickers et al. (2005), adapted with permission from the Endocrine Society (left); plotted from Gluckman et al. (2007c) (right).

suffer the consequences of doing so later in its life. However, a parallel induction of predictive responses also occurs in such situations.

Some studies have also looked at specific nutrient exposures. The level of vitamin A in a pregnant rat's diet influences the number of nephrons in her offspring's kidneys (Lelievre-Pegorier et al., 1998). The relationship between the level of vitamin A in the diet and the number of nephrons formed is linear. Determination of nephron number only occurs within a sensitive period in development before birth, well before the kidney replaces the placenta as the body's clearance system. Vitamin A may be the proxy for predicting later nutritional availability. Mature mammals usually have excess renal reserve. If a poor nutritional environment is predicted, then investing in excess and energetically expensive nephrons of the kidney may create a subsequent disadvantage in terms of survival and reproductive success, whereas such energy is best invested in early reproduction, as a long life cannot be anticipated. Conversely, in a rich nutritional environment, limited renal capacity is disadvantageous and leads to a damaging increase in glomerular blood flow, and consequently to an inability to adequately excrete metabolites and an increase in blood pressure.

The human data on 'programming' (a term we dislike because of the misleading nature of the metaphor, but extensively used in the literature), or DOHaD (as defined in Chapter 4), represent a continuation of this line of observation and thought (Gluckman et al., 2007a). The origins of this research were discussed in Chapter 4. A debate has centred around whether or not this form of plasticity reflects an adaptive process (Spencer et al., 2006; Wells, 2006; Godfrey et al., 2010). Interactions between parents and offspring will differ greatly according to the life history strategy of the species (Marshall and Uller, 2007). The interactions depend on whether the species is a high or low reproducer and the level of parental investment after conception. Oviparous and viviparous species have different intergenerational interactions in the embryonic/fetal stage of development. Species with high reproductive rates and no subsequent parental investment in their offspring, such as insects or many fish, will give primacy under limiting conditions to subsequent reproductive opportunities. Species, such as humans, that are low reproducers and show high parental investment, appear to give primacy to further investment in a pregnancy once established.

A human individual with a small body derived from a mother on a low nutritional plane benefits if, in a poorer environment, he or she is more likely to lay down fat and to seek out sources of high energy food compared with an individual with a large body derived from a mother on a higher nutritional plane. The first individual has a preference for high-fat foods and has a higher set-point for satiety and, we argue, has a phenotype that is more appropriate for survival in a poor environment. This functional argument is supported by evidence that the individual with a small body and metabolism to match does better than the extreme alternative in a poor environment. Children born smaller are less likely, in a famine, to develop kwashiorkor, the form of infant malnutrition with high mortality that involves a lower ability to mobilise substrates. In contrast, the low-birth-weight children respond to severe undernutrition by developing marasmus, which has much lower mortality (Jahoor et al., 2006). Large babies are more likely than smaller babies to develop rickets in famine conditions (Chali et al., 1998). In order for one phenotype to have utility in a particular environment, all that is necessary is that it be fitter than the alternatives under those conditions (see Figure 6.4). The conditions under which Darwinian selection of such anticipatory mechanisms have been extensively modelled are discussed in the next chapter.

These extensive studies of the effects of changes in maternal nutrition on the offspring's outcome have shown how maternal

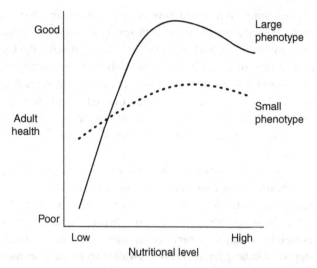

Figure 6.4. The hypothetical relationship between adult health and
nutritional level during later development for two extreme human
phenotypes (large, solid line; small, dashed line) that were initiated by
cues received by the fetus. From Bateson et al. (2004), reprinted by
permission from Macmillan Publishers Ltd.

condition affects body composition, metabolic control, neuronal
reserve, reproductive maturation, and behaviour. Generally, these
effects may be interpreted as being appropriate responses to the anticipated future environment (Gluckman et al., 2007a) with or without
associated immediate responses. The individual benefits by adjusting
the trajectory of its development so that its phenotype is most likely to
match the anticipated environment. In general, cues suggesting an
environment with scarce resources lead to a phenotype in which less
is invested in growth of the body. To support this accelerated maturation and to favour energy capture when available, the individual
develops a bias towards insulin resistance, thereby capturing the
higher-energy fat-dense foods when they are available (Gluckman
et al., 2010).

This interpretation of the evidence has been criticised, most
notably by Jonathan Wells (2006). The burden of his criticism is that
such a mechanism is unlikely to be adaptive in such a long-lived species
as humans. However, members of the hominin lineage have manifestly
been migratory, travelling to vastly different climatic regions of the
globe. Any mechanism that protected individuals from relatively short-
term changes in living conditions that differed from those in which

previous generations lived would have been highly advantageous. If a mother could transmit to her unborn offspring cues that affected its stature, metabolism and a host of life history characteristics, she would be at an advantage in terms of her own reproductive success over a mother who could not do this. Furthermore, as human reproductive success is largely dependent on survival through childhood (Jones, 2009), adaptations that affect such survival are likely to be found early in the life cycle, and that indeed is the case (Gluckman et al., 2010; Godfrey et al., 2010).

Part of Wells' argument is that, ultimately, the mother is in control. If conditions are dire, she can spontaneously abort her fetus or abandon her infant. On a more subtle level, she can trade off the benefits of fully supporting her current offspring against those of holding herself in readiness to produce another. This famous principle, first developed by Robert Trivers (1974), is widely accepted by evolutionary biologists and led to the idea of a conflict between the interests of parent and offspring. Yet the conflict principle can easily be overstated and often is. Empirical studies of parent–offspring relationships in mammals suggest that both parties adjust their responses to the state of the other and an equilibrium is reached. Two-way communication between mother and offspring benefits both sides (Bateson, 1994). A key issue is whether it is possible to distinguish between maternal weather-forecasting and maternal manipulation. The acid test is whether the small baby will be better suited to the poor environment predicted by the mother's low nutritional level than a big baby. We argue that it is. Further evidence against the maternal manipulation argument is provided by the observation that the antenatal environment can have long-term effects on the offspring without any shift in the neonatal phenotype (Gale et al., 2006). In such cases any maternal benefit is simply derived from the advantages derived by the offspring.

While the functional explanations of responses to cues in the fetal environment are well able to handle much of the data obtained from humans, it may not be appropriate for cases where a baby is born to an extremely obese mother. The adverse consequences for the child may represent something that is entirely new in the history of the human species, and may be a condition to which the fetus has no adapted response.

If fitness-protecting arguments are advanced, supportive evidence might be expected from studies of reproductive biology. In humans, menarche occurs earlier in hunter-gatherer societies in which juvenile mortality is higher (Walker et al., 2006). Children who are born smaller also tend to have earlier puberty (Sloboda et al., 2007); this can be particularly accelerated where the lower birth weight is associated with fat gain in childhood.

Similarly in modern societies, when a girl is exposed to stressful conditions in early life such as being orphaned, or the separation of her parents, she is likely to enter puberty at an earlier age than would otherwise have been the case (Belsky et al., 1991). She may express a strong wish to have children at an early age and become a mother in her mid-teens. This is commonly seen as a bad outcome for the young woman, at least in a modern society. However, the potential biological benefit is that she has a child while she is still able to do so. If bad conditions in early life forecast bad conditions later on, her best chance of increasing her fitness is to breed early (Nettle et al., 2010). The explanation seems plausible, but it is also important to separate the phenomenon with which it deals from another one that also involves early menarche; namely, the effects of good nutrition in early life.

LEARNING

Filial imprinting in birds and mammals, which occurs early after birth or hatching, and which is remarkably stable after it has occurred, has clear benefits in terms of recognising close kin. The onset of the sensitive period for imprinting is accompanied by other developmental changes that make it much easier for the young animal to learn about the details of its mother's physical characteristics. The young animal's vision improves around this time and it begins to move about with greater ease. The timing of this cluster of developmental changes depends on the species in question (Bateson, 1987). In birds that are hatched blind, naked and helpless, such as swallows, the onset of imprinting occurs much later in relation to hatching than it does in the precocious ducklings, which are already feathered and active when they hatch. The timing of behavioural imprinting can, within limits, be adapted to circumstances. Exposure to one object leads the individual to prefer it and reject anything perceived as different. Thus, experience of one sort prevents other types of experience from having the same impact. Such competitive exclusion is similar to what happens to visual development when one eye is covered up in early life. If the developing animal is deprived of the experience it would normally receive during the sensitive period, the imprinting process slows down and the sensitive period is lengthened. Hence, the sensitive period for imprinting is extended in domestic chicks if they are reared in isolation from other chicks. This flexibility of the imprinting mechanism is important because it allows for naturally occurring variations in conditions. When the weather is warm, for example, the mother duck will lead her ducklings away from the nest within hours of hatching. When the weather is cold, however, the mother will brood her young for several days after they have hatched and the young may consequently see little of her until then because they are underneath her. The benefits of having flexible neural machinery responsible for imprinting are clear in that it enables the young animal to cope with variability in the conditions of the environment.

Many other stable preferences and habits can be established by learning early in life. Making fundamental changes to mature behaviour patterns or personality traits may take time, resources, and quite possibly the support of other members of the individual's social group. So, in this respect, strongly entrenched behaviour acquired by learning can be almost as difficult to change as morphological characters. Adults have important tasks to carry out, such as feeding and caring for their family, and cannot readily reconstruct their behaviour without others to care for them during the transition phase. Bodies and behaviour can sometimes be changed, particularly under chronic stress and exposure to a new set of formative conditions. The benefits of doing so can be great when the new conditions indicate that the nature of the environment has changed permanently (Bateson and Martin, 1999).

Unlike the forms of plasticity that occur early in development, much of the capacity to learn can occur throughout life. But like the forms of plasticity we discussed earlier in this chapter, the delay between the detection of the cues that change behaviour and the benefits of doing so can be of short or long duration. The regulation of learning which we discussed in the last chapter is clearly related to utility. In many cases, cause and effect are closely linked in time, and the benefits of only learning about an association when an initially neutral event is followed within a short period by one with biological significance are great. However, when a novel food is followed by sickness, it is advantageous to allow a much longer period of many hours when establishing an association that leads subsequently to persistent reluctance to eat such food.

When learning does not occur after one exposure, the response to an environmental challenge may be gradually strengthened by learning when appropriate, or it may be gradually weakened when it is inappropriate. Nevertheless, animals learn much more efficiently when weakening occurs at a lower rate than strengthening (Wagner, 1981).

The benefits of doing so relate to the structure of the environment where, for example, the information carried by a reward is greater than that carried by the lack of one.

MISMATCHES

The loss of fitness that occurs when an individual's phenotype does not match the conditions to which it is adapted provides powerful evidence for the current utility of the character when a match does occur. In the case of the East African grasshopper, its greenish colour provides good camouflage against predators before the grass is burned by a savannah fire but is highly disadvantageous after a fire (Rowell, 1971). The black cuticle of the alternative form is beneficial against a blackened background, but after the rains when the grass grows again, it becomes costly to the individual. Such examples provide good tests of functional explanations (Figure 6.5).

The human eyeball grows in depth during childhood. The degree of growth relates to the visual exposure of the child such that the rate of growth is proportionate to, and controlled by, the point of focus on the retinal fovea. After childhood, eyeball growth is complete. Children reared in open spaces without a requirement to read and without exposure to artificial light rarely develop myopia (short-sightedness). In contrast, children reared in modern urban environments with much visual experience in artificial light and with a closer focus on written text are much more likely to develop myopia. Indeed, an association between the incidence of myopia and increased education has been observed (Milinski, 1999). Optimal distance vision is clearly linked to early childhood visual exposure to the horizon (Dirani et al., 2009) and in the modern urban world, where prolonged exposure to open spaces is limited, the mismatch between the biology of vision development and the visual environment of the present becomes manifest.

Some adaptations that benefit humans in relatively impoverished conditions may become dysfunctional in a radically different affluent world. This has been discussed extensively in relation to the phenomena clustered under the heading of DOHaD (see Chapter 4; Gluckman et al., 2010). Behaviour such as seeking out and eating fatty foods is vital in a subsistence environment, but in a well-fed society it does more harm than good to the individual. Fetuses exposed to famine in pregnancy themselves grow up to have a higher risk of coronary heart disease and obesity (Painter et al., 2005). Similarly, disordered lipid metabolism, appetite regulation and body fat

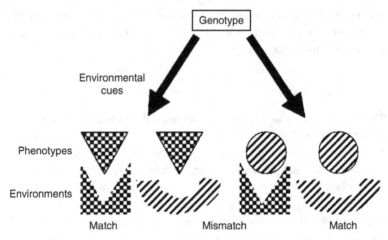

Figure 6.5. A given genotype may give rise to different phenotypes depending on the state of the environment early in development. Cues from the environment may be used a predictors, determining which of a set of alternative developmental pathways is elicited. If the environment does not change, then the organism's phenotype will be well adapted to that environment, providing a close match, as represented in the diagram by the pattern and shape of the phenotype and the pattern and shape of the environment. However, if the environment does change between the elicitation of the particular pattern and development, then the phenotype may be mismatched to the conditions of adult life. From Bateson (2007), American Society for Nutrition.

composition have been observed in rats undernourished *in utero* and these effects are exacerbated by a high-fat postnatal diet (Vickers et al., 2005; Erhuma et al., 2007; see Figure 6.3). To summarise, where the prediction of an adverse environment is wrong and the offspring lives in an enriched environment, then its physiology and environment are mismatched – a mismatch that brings with it an increased risk of metabolic and related disease in later life. This may not be early enough to compromise reproduction but may deny to grand-offspring the benefits of having grandparents. Where the offspring is predicted a threatening environment it may opt for early reproduction at the expense of longevity.

Such mismatches that can arise in human populations experiencing rapid improvements in their standard of living raise important issues for public health. They have attracted the unfortunate term 'programming' for the developmental processes giving rise to ill-health in later life. This term has been commonly applied only to the consequences of poor nutritional environment or maternal stress early in development. In any event, the term implies – from its roots in computer science – that instructions have been fed into the organism, when in our view the phenomenon merely suggests that the developmental trajectory has been elicited by an environmental cue (Bateson, 2007).

Gambling, which can so easily become compulsive and ruin a life, seems wholly irrational but makes psychological sense in a world in which the delivery of rewards is rarely random. If you have done something that produced a win, it is usually highly beneficial to repeat what you did – except when you are in a casino. Similarly, the tendency of parents to protect their children from all contact with unknown people after hearing about a child murder on television is beneficial in a small community where such news might represent real danger. In a modern context, where enormous prominence is often given to such tragedies, such risk-averse behaviour in Western societies – in which the incidence of child murder has remained constant for decades – merely impoverishes their child's development and may have longer-term psychosocial consequences. For example, reducing the opportunities for playing outside with other children can be highly detrimental to long-term development (Pellegrini, 2009). Some of the dire predictions are already being seen in rising levels of obesity in children and young adults.

CONCLUSIONS

Various mechanisms that generate robustness and are involved in plasticity coexist to allow the development of an integrated phenotype or a variety of alternative phenotypes. Successful lineages of sexually reproducing organisms require compatibility between the genomic architecture and the phenotypes at a number of levels. This requires a level of robustness in development. But organisms living within variable environments also require plasticity to cope with environmental change. To ensure their utility, these mechanisms must be integrated with those that maintain the species' characteristics.

The utility of a match between an individual's characteristics and the environmental conditions in which the individual finds itself is revealed when, for whatever reason, other individuals do not have the characteristics that match those conditions so well. If the individual without the feature in question is not less likely to survive or reproduce, then an evolutionary explanation for the current utility of that feature based on Darwinian selection is questionable. However, this does not mean that the trait itself does not have an evolved origin. As discussed in Chapter 2 it may, for example, reflect co-option of the trait for a different function. The features that lead to a well-regulated phenotype might have evolved in a number of different ways. These historical issues are discussed in the next chapter.

7

Evolution of developmental processes

Evolution is a fact. No serious biologist disputes that organisms have changed over time or that they continue to change. Other organisms have become extinct, many of them in recent times. What requires explanation is the way in which these changes take place. Darwin observed that members of a species differ from each other, that some were more likely to survive and reproduce than others, and finally that the characteristics of the fitter individuals would generally be inherited by their offspring. So, by the process he metaphorically termed 'natural selection', lineages would evolve. Like his contemporaries, Darwin knew nothing of the molecular processes of inheritance and was often tempted into supposing that acquired characters could be inherited, arguing for example that behaviour patterns that are learned generation after generation will eventually be expressed without being learned. In the mid-twentieth century, the genetic mode of inheritance which had become established at the beginning of the century was brought together with Darwin's evolutionary theory in what was called at the time the Modern Synthesis. This conceptual framework became the dominant mode of biological thought, emphatically reinforced by the culture of genetic determinism that has accompanied the explosion of genomic knowledge.

Evolutionary theory itself continues to evolve, integrating the views of the Modern Synthesis with the explosion of observation and theory coming from the developmental, ecological and molecular sciences (Pigliucci and Müller, 2010). Other mechanisms of inheritance have been discovered. These include transmission across generations through the transgenerational passage of symbiotic bacteria (Gilbert, 2005), transgenerational transmission through the Y chromosome (Nelson et al., 2010), direct epigenetic effects transmitted through meiosis and indirect epigenetic effects such as those transmitted through the mother's behaviour, and social learning (Jablonka and Raz, 2009). Moreover, the nature of developmental processes discussed in this book means that some forms of evolutionary change are more likely to occur than others. This awareness has spawned attempts at a new synthesis between theories of development and those of evolution – the evolutionary developmental biology movement or so-called 'evo-devo' (Gilbert et al., 1996; Amundson, 2005; Pigliucci and Müller, 2010).

A key purpose of this book has been to clarify concepts which, despite these attempts at a new synthesis, have remained obscure due to a lack of mechanistic explanation, insufficient clarity of terminology, or failure to provide a coherent model. Two fundamental issues have remained in bridging the gap between the neo-Darwinist camp and those who seek to emphasise the importance of development in evolution. The first has been the need to provide molecular mechanisms that would explain the role of development in evolutionary processes. The second has been the need to demonstrate the generality of developmental processes impacting on evolution. The growing understanding of epigenetic biology in the broadest sense addresses, to a considerable extent, the first of these, and in Chapter 8 we shall discuss examples that address the second. In this chapter we focus on the question of the evolution of developmental processes themselves. In earlier chapters we stressed how the component processes of development interact with each other in the development of the individual phenotype; in this chapter we shall examine how these processes generating robustness and plasticity might have evolved and how they are integrated, given the ecological conditions in which the organism exists.

ADDRESSING EVOLUTIONARY HYPOTHESES

Testable ideas about the current utility of a trait are often used to propose explanations of its evolutionary origins. Indeed, many of the explanations that we offered for the benefits of integrated development, which we discussed in the previous chapter, furnish suggestions for the way in which the mechanisms evolved. Nevertheless, some caution is needed, since such reasoning has its limitations and may be faulty. The benefits of the past are not necessarily those of the present. The adaptive advantage of a particular trait in earlier and differing circumstances need not relate to the significance, if any, of the trait in current conditions. The characteristics of the current trait may be different from those of the trait from which it originally evolved. For example, the bones of the mammalian middle ear have their evolutionary origin in the jawbone of fish. Even if a plausible historical explanation could be constructed, the feature under study may have been a side-effect of an evolutionary process – what Gould and Lewontin (1979) likened to an architectural 'spandrel'. In general, the inference from the function at one period in evolutionary time to another, either forward or back, can clearly be misleading.

In addition, historical explanations are, by their nature, difficult to test. Direct observation is impossible but, at best, multiple lines of inference can provide for a robust conclusion. For example, inferences drawn from comparisons between different populations of the same species, or between different taxonomic groups, can be compelling. Modern comparative biology allows a focus on the degree of difference between taxonomic groups with the aim of revealing their phylogenetic relatedness and pinpointing when the lineages diverged in the past. It may also answer whether or not the phenotypic similarities are due to convergences where a common problem set by the environment has been solved in analogous ways by unrelated lineages. In particular, our increasing understanding of molecular architecture and the potential to read molecular history in genomic sequences, and indeed to recover DNA from fossils, allow for an improved understanding of what might have happened. Sequencing of large components of the Neanderthal genome suggests that some hybridisation occurred between that species and some members of *Homo sapiens* in non-African populations (Green et al., 2010). As a result, there is a small Neanderthal contribution to the genomes of some modern humans.

Many well-documented instances exist where homologies reflect phylogenetic relationships. Examples include the relationship between the gill arches of the fish and the vascular anatomy of the thorax and neck of primates, and the relationship between the bony structure of fins in whales and the limbs of other mammals. Conversely, while the cephalopod and the vertebrate eye show remarkable similarities, the independence of their evolutionary origin is well demonstrated by the different anatomical relationships between neural innervations of the retina and the light-sensitive cell (Fernald, 2000). In vertebrates, innervations are from the front of the eye, requiring the optic nerve to pass through the retina, thereby creating a blind spot. In cephalopods, the optic nerve runs from the back of the retina – so there is no blind spot.

Another example of different evolutionary strategies achieving the same functional task is provided by two human populations both successfully living for many generations at altitudes above 3,000 metres, one in Tibet and the other in the South American Andes.

To live and reproduce at that altitude requires several adaptations to cope with the low atmospheric oxygen content. While both the Tibetan and the Andean populations have evolved this capacity, they do so in quite different ways (Beall, 2007). Andean highlanders have higher oxygen levels in arterial blood, and higher levels of the oxygen-carrying protein haemoglobin and the inducer of its production, erythropoietin. In contrast, Tibetan highlanders compensate for hypoxia through higher resting ventilation and pulmonary blood flow and do not have high haemoglobin concentrations. Genetic analysis has revealed that several variants of the gene *EPAS1*, which encodes a protein that regulates haemoglobin levels in the blood, are found at high frequencies in the Tibetan population (Beall et al., 2010). These variants strongly correlate with lower haemoglobin concentration, which is thought to protect against altitude sickness. The correlation does not establish a causal relationship but does suggest that *EPAS1* has been spread in the population as a result of the greater fitness of those who deal best with high-altitude hypoxia.

MODULARITY

A problem for the evolution of a well-integrated organism is how to change one part without it having ramifying effects on all the other parts. One solution is to evolve modularity. A system is modular if it is composed of multiple parts, each of which is tightly integrated within itself and operates with a certain degree of independence from the other parts. The quasi-independent parts are the modules, and the greater their independence from each other, the more modular the system. A phenotypic trait is developmentally modular if the processes responsible for its development are quasi-independent of the processes responsible for the development of other traits in the same organism. Developmental modularity is now the subject of intense study by evolutionary and developmental biologists, expanding upon a suggestion by Lewontin (1978). Developmental modularity may well be an important condition for rapid and efficient evolution when environmental conditions change rapidly (Wagner and Altenberg, 1996).

Segregation of characteristics in closely related individuals and independent inheritance of some of the factors necessary for development are commonplace. The shuffling of discrete and supposedly inherited characteristics from one generation to the next is a feature of sexual reproduction. In human families it is obvious that a child's characteristics are not a simple blend of its parents' characteristics.

Most parents will find some particular likeness between themselves and their child. A daughter might have her mother's hair colour and her father's shyness, for instance. The child may also have characteristics found in neither parent: a son might have the jaw of his grandmother and the moodiness of his cousin. In the breeding of animals, artificial selection for behavioural characteristics such as tameness can be extremely rapid (Trut et al., 2009).

Findings such as these have suggested that different elements of the genome are quasi-independent in relation to their effects on developmental processes and phenotypic outcomes. In terms of their evolution, modules might have arisen through biased mutation or they might have arisen through the benefits they provide to a lineage (Wagner et al., 2007). The two possibilities are not mutually exclusive, since one might have led to the other.

EVOLUTION OF ROBUSTNESS

In Chapter 3 we identified multiple ways in which robustness could be conferred on development. Some of these ways were almost certainly not the products of positive Darwinian selection. For example, robustness due to insensitivity to the environment might simply reflect a species having been stable in an unchanging ecological niche over evolutionary time; plasticity would have had relatively little utility. Alternatively, where sensitivity to environmental cues might induce harm, mechanisms to render the developing organism insensitive to the environment may be under active Darwinian selection. Viviparity in some reptiles and fish, the firm eggshell of birds, and the placental barrier of mammals all represent evolved systems to create environmental insensitivity (among other potential advantages).

Some constraints on plasticity may have had little or no evolutionary advantage but may merely have been the by-product of the lack of Darwinian selection for plasticity in a specialist species; alternatively they may have evolved because of the interaction with other traits, or because of the limitations imposed by other traits. While gestational length is somewhat plastic in many mammals, it is effectively constrained by the deficits of prematurity at one extreme and by placental physiology and the size of the maternal pelvic canal at the other (Gluckman and Hanson, 2004). The possible energetic costs of maintaining plasticity may have created further constraints on the capacity of the organism to respond to change. Finally, robustness generated by an attractor in a complex system (see Chapter 3) is likely to be a secondary consequence of the evolved benefits of complexity. On the other hand, robustness due to repair mechanisms, and the presence of redundant systems, are likely to have been the products of Darwinian evolution.

In some species the potential for complex repair can be high. Salamanders can regrow lost limbs due to complex mechanisms that include the capacity for mesenchymal cell de-differentiation at the stump followed by fibroblast growth factor (FGF)-stimulated innervation and thus regrowth. In other amphibians such as frogs, the FGF signalling pathway loses functionality upon metamorphosis into the adult form and the ability to regenerate limbs is lost after this stage. In evolutionary terms the presumption is that the cost of maintaining such a capacity is high, and so it has only been sustained in phyla and species where the risk of losing a limb is sufficiently high.

Redundancy in biochemical processes and in anatomical and cellular components reflects other evolved means of ensuring a robust phenotype, and such redundancy abounds in multicellular organisms (see Chapter 3). These mechanisms all operate to produce organisms that have the capacity to maintain a wide variety of their characteristics. Presumably the evolutionary advantage to organisms in possessing such systems would have been the ability to cope more satisfactorily with an array of environmental challenges and some degree of injury than organisms without them.

ROBUSTNESS AT THE MOLECULAR LEVEL

Most theories about the origins of life hold that the earliest organic mechanisms of inheritance depended on RNAs that provided the replicating sources of information. However, RNA does not generally have a paired structure, so errors of replication are highly likely. The emergence of double-stranded DNA as the primary replicator is considered to be one of the major transitions in evolution (Maynard Smith and Szathmáry, 1995); its duplex structure allows high fidelity during replication by providing a back-up template that enables restoration of the original sequence should a mutation occur on one of the strands (see Chapter 3). In addition, DNA replication involves several other proofreading mechanisms such as an exonuclease that immediately detects and cleaves incorrect nucleotides during polymerisation, and mismatch repair enzyme activity that occurs post-replication (Polosina and Cupples, 2010). As a result, consistency in genomic information can be achieved between generations. This stability was an essential element in creating the phenotypic diversity on which Darwinian selection could act. The phenotypic consistency of asexually reproducing eukaryotes and transmitotic phenotypic reliability in multicellular organisms reflect such stability. Similar processes operate at meiosis. Consequently, even though recombination events lead to allelic switching in the chromosome, the mismatch repair processes are generally able to intercept and correct mismatched bases and largely maintain genetic fidelity.

The integrity of the genome can be challenged in ways other than through inaccuracies occurring during replication. After invasion of a virus, the organism's transcriptional and translational machinery are hijacked to produce viral nucleic acids and proteins. In the case of the RNA-based retroviruses, the enzyme reverse transcriptase is used to manufacture complementary viral DNA from its RNA genome, and the viral DNA is then incorporated into the genome of the organism (Kurth and Bannert, 2010). If the site of insertion contains a protein-coding or regulatory region, gene expression may be altered and normal functioning is potentially subverted. Retroviruses such as human immunodeficiency virus target somatic cells; those that invade the germline, such as the family of human endogenous retroviruses that have been implicated in some cancers, can transmit their DNA to the next generation, thus persisting in the host genome. Retroviruses are related to retrotransposons, which, with other transposable elements, comprise as much as 40% of the mammalian genome (Goodier and Kazazian, 2008). Some transposable elements have been linked to genetic diseases in humans (Wallace et al., 1991).

The majority of transposable elements are not expressed, although they have provided additional DNA template to the host (Kidwell and Lisch, 2001). Initial silencing by epigenetic mechanisms probably occurred until mutations accumulated and the sequence could no longer be expressed. Indeed such epigenetic silencing probably evolved either as the key process for the initial silencing of viral infection or for silencing the second genome in the case of cell fusion (Grandjean et al., 1998). Possibly the robustness of the host's phenotype was maintained by the evolution of mechanisms that silenced transposable elements.

However, many transposable elements have had evolutionary effects. The inserted sequences provided new DNA sequence material that could benefit the recipient, and this was likely to be important in the emergent complexity of gene regulation across phylogeny. For example, the gene coding for a cell-surface glycoprotein in humans was inactivated by a retrotransposon event after divergence from the last common ancestor with chimpanzees about 2–3 million years ago (Chou et al., 2002). The different cell-surface properties that resulted are linked to human resistance to infection by the chimpanzee malaria parasite *Plasmodium reichenowi* and, conversely, chimpanzee resistance to the human-specific species, *P. falciparum*. Resistance to *P. reichenowi* might have been the evolutionary mechanism that led to fixation of the retrotransposon-induced mutation (Wang et al., 2003).

Gene duplications offer another form of robustness in that they reduce the likelihood of a mutational effect if one of the copies is affected (Wagner, 2008). Moreover, gene duplications offer the opportunity for the evolution of diversification, with the original gene maintaining its function and the copy taking on another function (Wagner, 2008). The human insulin and insulin-like growth factor (IGF) 1 and 2 genes are all evolutionary duplicates of an ancestral insulin-like gene: while the three gene products have some overlap of function, their primary functions differ, with insulin regulating glucose uptake and the IGFs regulating cell growth.

The evolution of mechanisms that constrain developmental plasticity through promoting canalisation would have played important roles in the evolution of robust phenotypes. The heat shock protein gene *Hsp90* is found in many taxa; its gene product maintains the conformational stability of proteins during synthesis and heat exposure, and is also thought to act as a buffer against genetic variation (Rutherford and Lindquist, 1998). In *Drosophila* with a mutation at the *Hsp90* locus, considerable variation in morphological aspects of the phenotype, such as deformed legs and discoloured eyes, is observed. The presence of this phenotypic variation in the population suggests that *Hsp90* acts to conceal variation which is only revealed when the gene has been rendered functionless by mutation and the developmental process has been 'decanalised' (Sgrò et al., 2010). Sollars et al. (2003) suggested that *Hsp90* is associated with epigenetic mechanisms. Indeed, Gangaraju et al. (2010) found that a piRNA-binding protein known as Piwi is involved in canalisation and is regulated by *Hsp90*.

EVOLUTION OF PLASTICITY

In Chapter 4 we considered several distinct types of plasticity at different levels of organisation. For example, the individual born with deafness accommodates through increased use of other modes of communication and is able to survive and reproduce. A variety of birth defects such as cleft palate or limb abnormalities

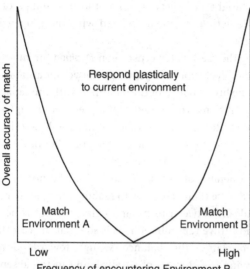

Figure 7.1. The conditions under which plasticity was likely to have evolved, when the environment might be in one of two states: A or B. If the frequency of occurrence of either environment (A or B) over evolutionary history has been high, then plasticity is less likely to evolve. However if the environment had tended to vary between A and B then plasticity would be more likely to evolve.

can frequently be accommodated, even in species that do not have the social support found in humans.

Other mechanisms described in Chapter 4 adjust the individual's phenotype to local environmental conditions, either those currently experienced or those anticipated in the future. The East African grasshopper deposits black melanin in its cuticle if the reflectance of the ground is low when it hatches out, as it would be after a savannah fire. As a consequence of this mechanism, most grasshoppers are green in periods without fires and most are black in periods after fires when the savannah is blackened. The question therefore arises: under what conditions would the plasticity of the grasshopper evolve? If fires were infrequent and quickly followed by rain, it might be disadvantageous ever to be black. Conversely, if fires were frequent it might be advantageous to remain black at all times. In between these two extreme conditions, it would be highly advantageous to be capable of matching colour to the background (Figure 7.1). Whether or not the plasticity evolved would depend on whether individuals appeared that were capable of making the switch. If plasticity appeared and spread through the population, it might disappear again if the probability of fires dropped and an energetic cost was associated with the propensity to be plastic.

In the case of the freshwater crustacean *Daphnia*, mentioned in Chapter 4, Darwinian evolution has provided the young animals, still developing within the brood pouch of their mother, with the capacity to anticipate future conditions. The presence of a predatory midge in the water causes the young to form a defensive helmet and long tail spine. In the absence of the midge (or, more accurately, the chemical remains of *Daphnia* killed by the midge), the young do not develop the armour (see Figure 4.2). The benefit of not doing so is that the non-helmeted females are able to devote their resources to making many more eggs in adulthood. The trade-off between forming a helmet or not is between present survival and future reproduction, as is so often the case. In the presence of the predator, the balance swings towards devoting resources to improving the chances of survival, and in the absence of the predator the balance swings the other way towards producing more offspring. The maintenance of the capacity for such flexibility will depend on historical conditions. If predators had always been present and the capacity for changing the course of development carried a cost, this capacity would almost certainly have been lost. Conversely, it would almost certainly have been lost if predators that could be deterred by armour were never present.

In Chapter 6 we described the effects of differing day length on the pregnant Pennsylvanian meadow vole. Coat thickness in the off-spring is dependent on whether the mother experiences lengthening or shortening periods of daylight during pregnancy. The mechanism is plausibly an evolved adaptation arising from correlated seasonal changes in temperature. Those mothers that did not signal the future conditions of the environment to their unborn offspring would have had a lower reproductive success than those mothers that did. Correspondingly, those offspring that failed to respond to the maternal cue would have been less likely to survive after birth than the respon-sive ones.

Where the cue is regular, such as seasonal change, it will have high fidelity. Selecting an appropriate developmental trajectory in response to the cue carries little risk and much advantage. When the cue reliability is less than perfect, the evolutionary benefit of a plastic response to environmental conditions must be greater than the cost of producing an inappropriate phenotype. Here is a rich seam for the theoreticians to mine (Moran, 1992; Lachmann and Jablonka, 1996;

Sultan and Spencer, 2002). Such anticipatory responses might confer an advantage in terms of survival and reproductive success, not simply where the environmental variation is seasonal or stochastic, but also where the effect of the environmental conditions can be averaged over the life course of the parent or even over several generations (Kuzawa, 2005). If that were the case, a gradual change in phenotypic characteristics could be smoothed over generations as environmental conditions slowly changed.

In considering human pregnancies, these considerations have led to the development of the concept of inertia in the response to a nutritional cue during development, whereby the fetus does not show significant response in plastic terms to the minor day-to-day fluctuations in maternal intakes; instead it responds to changes in maternal nutrition as reflected in body composition over a longer time base (Kuzawa, 2005). Similarly, the placental barrier to maternal cortisol largely buffers the fetus from transient stress-related shifts in the mother.

While inappropriate prediction is a risk, modelling suggests a selective advantage under these conditions even when the probabilities of accurate prediction are marginal (Jablonka et al., 1995). The key assumption is that the predictive plastic response confers some advantage on the offspring in terms of survival. This will certainly be so where the inducing response is subtle or where the species is a slow reproducer and parental investment in the offspring is high.

However, in species that reproduce frequently, maternal interests are more likely to have primacy, and the offspring is sacrificed or compromised to protect the maternal interests for subsequent reproduction in more favourable conditions. The situations in which maternal advantage might be favoured have been subject to theoretical analysis (Marshall and Uller, 2007) and were discussed in Chapter 6.

A further theoretical consideration arises because the risks of making an incorrect decision are often asymmetrical. An individual adapted to environment A but ending up in environment B may be less severely affected than an individual adapted to environment B ending up in environment A (Moran, 1992; Sultan and Spencer, 2002). For example, a human that as a fetus is set on a trajectory towards a phenotype adapted to a poor nutritional environment in later life, but then ends up in a rich environment, may suffer little or no health damage during childhood and through the reproductive phase, but may experience deleterious effects later in life with little effect on his or her fitness (Gluckman et al., 2005b). By contrast, an individual with a phenotype adapted to a highly nutritious environment but ending up in a poor one may be severely affected – his or her nutritional requirements might be more rapidly compromised in famine, and in women successful reproduction becomes impossible (Chali et al., 1998; Eriksson et al., 2003). This asymmetry may explain why children of larger birth size are more likely than those of lower birth size to develop kwashiorkor, a potentially fatal form of infant malnutrition, in famine conditions (Jahoor et al., 2006). The general point is that a human who is mildly or moderately obese is generally healthy through the reproductive years. Therefore to predict a poor postnatal environment and live in a high-nutrition environment may be less deleterious to reproductive success than developing a body that requires high energy support and ending up in a poor nutritional environment.

The delay between the detection of an inductive cue and the full expression of the phenotype is sometimes lengthy, as in the human case. A complex body cannot be built in a trice and adaptations to particular environmental conditions are often complex. This lag explains, in part, why a phenotype – once developed – cannot be readily changed and, as we have observed, a mismatch of phenotype to environment can therefore arise if the forecast of local conditions proves incorrect. A further conceptual issue concerns the optimal time lag between detecting a cue that predicts a given set of environmental conditions and the phenotypic response to that cue. A hasty response might mean that conditions could change again before the adaptation becomes relevant. Left too late, and the capacity for plastic change might be exhausted or the adaptation may not be developed in time to be effective. The case of the East African grasshopper is instructive here because the process of developing a given phenotype can be reversed early in development but becomes progressively more difficult at each successive instar stage, presumably reflecting the developmental constraints on body construction discussed in Chapter 3. The theoretical issues related to this issue of the temporal relationship between cue and response have been considered elsewhere (Nishimura, 2006; Wake et al., 2010).

Some species are better able to cope with variable environments than others. The former are often termed generalist species, whereas the latter, known as specialist species, are often limited to rather narrow but stable ecological niches; extinctions are often associated with the loss of such specialist habitats. The capacity to cope with variable environments may be inherent in the evolved adaptive physiology of generalist organisms, reflecting the range of environments that the organism and its ancestors may have experienced over evolutionary time. Humans, for example, can cope with a wide range of foods and macronutrient balances. Some humans are exclusively vegetarian and others have a high meat intake, but koalas and pandas can only survive on particular types of vegetation. Alternatively, or additionally, the capacity to cope with variable environments over time may depend on developmental plasticity. This plasticity may be manifested as polyphenisms, where the phenotype that has particular advantage in one environment is quite distinct from that which has advantage in a different environment; or the developmentally induced reaction norm may be continuous.

Developmental plasticity may confer the potential to cope with a wider range of environments than would otherwise be possible, and also to sustain fitness when environmental conditions fluctuate, particularly when the environment changes relatively slowly. When environments remain constant over long periods of time, the benefits of developmental plasticity are lost. The likelihood of loss would become greater if the underlying mechanisms of developmental plasticity were energetically costly to maintain.

Many of these principles can be well demonstrated in plants, where parental effects have been particularly well documented. Sultan et al. (2009) studied two ecologically distinct but closely related species of annual plants: a generalist that could cope well with both dry and moist conditions, and a specialist restricted to moist conditions. They demonstrated that offspring of the generalist species showed adaptive and plastic responses to drought, such as larger root systems, which were not found in the offspring of the species that was specialised to live exclusively in a moist environment.

Many organisms are characterised by having complex life cycles, with each stage involving a quite distinct phenotype. Switches between sexual and asexual forms may occur, or serial morphs may occur within one life cycle, as in the case of the Lepidoptera or Amphibia. Complex life cycles may involve different forms in different environments, as in the case of a number of parasites such as the malaria-causing protozoan *Plasmodium*, which exists as various morphs and in both asexual and sexual forms depending on the stage of the life cycle and whether the host is a mosquito or a vertebrate (Koyama et al., 2009). Each of these forms is triggered by developmental switches, and in some cases environmental cues play a major role. Minelli and Fusco (2010) have pointed out the similarities in principle between the evolution of such complex life cycles and the evolution of polyphenisms.

EVOLUTION OF LEARNING AND MEMORY

Some forms of plasticity involve rapid responses. Learning processes in all their different manifestations provide the obvious examples. The adaptive advantages of learning and memory confer additional capacity to recognise and avoid predation, to identify nutritional resources, to undertake sexual reproduction, and so on. Costs may arise, however, in terms of the time or energy required to learn a task – an issue that we return to in Chapter 8. The evolution of learning mechanisms evidently began early in the history of life and may be uncovered by a comparative approach. Edward Thorndike's monograph on animal intelligence began by explaining that the primary reason for studying learning in animals was to seek the roots of complex processes and, in so doing, to trace the evolution of learning (Thorndike, 1898, pp. 2–3). He hoped to provide an evolutionary framework for a true comparative study of learning. This part of his project was largely forgotten but was picked up again by Moore (2004) who argued that similar-looking processes such as behavioural imprinting evolved many times. He provided several examples of how more complex examples of learning might have evolved from less complex processes.

The process of sensitisation to environmental conditions, described in Chapter 4, is found in organisms that are extremely ancient. Even this form of plasticity found in simple organisms requires the ability to categorise information provided by the environment. A process similar to behavioural imprinting in birds has been described in the nematode worm *Caenorhabditis elegans* (Remy and Hobert, 2005). The worms respond preferentially to food odours to which they have been exposed in early life. With such a capacity to distinguish between different types of stimuli in place, a plausible case can be made for the evolution of associative learning from sensitisation (Wells, 1967; Kandel and Schwartz, 1982). Sensitisation, which is found in many simple organisms, involves an enhancement of responsiveness to an aversive stimulus, but without any association occurring between the aversive stimulus and the conditions that predicted its occurrence. If the sensitisation persisted over a long period then the evolution of a capacity to associate neutral and biologically meaningful stimuli, typical of classical conditioning, is readily understood. Ginsburg and Jablonka (2007) have argued that the evolution of associative learning then triggered the explosive radiation of evolution found in the Cambrian. Their argument is that an evolved capacity to predict and control the environment, which is the central feature of associative learning, provided such a major benefit that animals were able to exploit niches previously unavailable to their ancestors. The potential role of behaviour in driving evolution is discussed in the next chapter.

The evolution of immunological memory requires a different explanation from the evolution of behavioural memory. The explanation is not hard to find, since parasitism has been such a potent driver of evolutionary change. The costs to health of an immune system suppressed by drugs or chronic stress are so marked that the advantage of having an unimpaired one over not having one need hardly be argued. Nevertheless, the costs of running an immune system are substantial. Suppression of the human immune system during periods of chronic stress is often thought to be potentially harmful, leading to increased risks of infection and the growth of some malignant tumours (Smyth et al., 2006). In contrast, an evolutionary perspective views the immune system as an energetically costly system that may or may not have priority over other uses of that energy. From this perspective, the immune system may have energy made available for it via the reduction of other activities or it may be suppressed when other energy-consuming activities are more important than immunity for immediate survival (Segerstrom, 2007).

The widespread occurrence of plastic responses across the plant and animal kingdoms suggests that these processes generally confer great advantages. The magnitude and nature of the plastic response do depend, however, on environmental conditions (reviewed in Thompson, 1991; Dejong, 1995). An organism that demonstrates little plasticity yet maintains a high performance in optimal conditions may be favoured over an organism that is highly plastic yet only benefits from its plasticity in poor conditions. Social learning, whereby an individual acquires information about the environment from the experience of others, may be beneficial in many circumstances but can carry costs for the individual when it leads to sub-optimal performance by the whole group (Giraldeau et al., 2002). Such cases could arise, for example, when one individual makes a mistake and other members of its social group copy what it has done.

PLASTICITY AT THE MOLECULAR LEVEL

Plastic mechanisms probably evolved early in the history of life. Evidence for epigenetic regulation in bacteria was suggested by the protoplast fusion between strains of *Bacillus subtilis*, leading to the suppression of one genome (Grandjean et al., 1998). In the bacterium

Escherichia coli, the *lac* operon consists of a set of coordinated genes that act as a bistable, all-or-none switch involved in lactose metabolism; the presence of lactose inhibits production of a repressor protein and permits transcription of β-galactosidase, which catabolises lactose. Mathematical modelling has been performed with multiple *lac* operon variants, each possessing different point mutations at the regulatory site. Instead of functional disruption, the mathematical models suggested that this operon demonstrates plasticity by being able to integrate variable inputs in novel ways (Mayo et al., 2006).

In multicellular organisms, epigenetic mechanisms are involved in many developmental processes such as cell differentiation, gene dosage regulation, regulation of the expression from genes with multiple copy number, and genomic imprinting. In general, the epigenetic processes, though co-opted for many different functions, are highly conserved through evolution. The specific molecular epigenetic mechanisms used vary across species and taxa. For example, in honeybees methylation largely occurs within exons, whereas in mammals methylation is generally within promoter and non-coding regions of the DNA (Feng et al., 2010). *Drosophila* appears to have largely lost the capacity to methylate CpG sequences even though this ability is present in other insects; instead it seems to utilise other epigenetic processes including small non-coding RNA activity and histone modifications. Epigenetic processes may also have specific effects in regulating gene dosage in particular situations. For example, in the experiments on maternal licking by rats, altered methylation of a gene was involved in certain forms of stress response (Weaver et al., 2004), but the effect is graded (Weaver et al., 2007).

Histones may have evolved as a means of regulating gene expression through their role in modulating DNA packaging into chromatin, with epigenetic control points emerging as a secondary benefit (Felsenfeld and Groudine, 2003). As alluded to earlier in this chapter, epigenetic mechanisms, including RNA-mediated silencing, DNA methylation and genomic imprinting, have been postulated to have initially evolved as defence mechanisms to silence transposons and protect the genome (Slotkin and Martienssen, 2007). An additional argument has been advanced in the case of plants and early vertebrates, proposing that transposon insertion during evolution invoked host genome defence mechanisms, leading to epigenetic regulation by promoting methylation (Matzke et al., 2000).

Epigenetic modifications may provide a stable basis for memory, as suggested many years ago by Griffiths and Mahler (1969) and supported by subsequent empirical work. Neuron-specific overexpression of HDAC2, a histone deacetylase that is involved in chromatin remodelling, impairs various physiological indices of cognitive function and memory formation in mice. As predicted, the administration of histone deacetylase inhibitors could ameliorate some of the cognitive loss, and indeed HDAC2 appears to associate with the promoters of genes relevant to learning and memory (Guan et al., 2009). Further recent work has implicated histone methylation in memory formation, and DNA methylation in memory maintenance (Miller et al., 2010; Peleg et al., 2010). This growing body of knowledge may be a further demonstration of how an epigenetic mechanism that originally evolved to protect the genome from disruption was co-opted to enable the organism to gather and use important information from its environment.

CONCLUSIONS

The evolution of a successful lineage at the level of the whole organism depends on having a robust phenotype that is well matched to the environment. However, if the environment is variable, having the capacity for plasticity becomes advantageous. The extent of the plasticity is likely to depend on the level of variation, its frequency, and the capacity of homeostatic mechanisms to cope with short-term variation. Some forms of response cannot be adequately met by changing homeostatic set-points because of structural and energetic considerations, or because the types of environments experienced are so extreme that an integrated but distinct phenotype is best achieved through a discontinuous range of body types – that is, polyphenism – as in the case of the desert locust. Indeed metamorphosis can be seen as a serial form of polyphenism within a single life course, and the timing of metamorphosis, as we have seen, can indeed be considered plastic.

In this chapter we discussed the evolution of the various components of robustness and plasticity. Separating out the components might seem to conflict with the overall theme of this book, which is that these components are integrated in the development of an individual organism. We think not. The components of robustness and plasticity almost certainly evolved at different stages in the past; some, such as DNA repair enzymes, are present in prokaryotic organisms; whereas others, such as tissue repair mechanisms, are features of multicellular organisms. Moreover their inheritance can be modular. The evidence does imply that the independent inheritance of some of the factors necessary for development happens all the time. Once in place in a given organism, the integration into a phenotype that best serves the survival and reproductive needs of the organism will depend on the particular ecological conditions in which that lineage is evolving. In the language of 'evo-devo', the components form the toolkit from which the phenotype is fashioned. In the next chapter we examine how the phenotype can then play a role in the evolution of the organism's descendants.

8

Impact of developmental processes on evolution

For many years the debate about the role of an individual's development and its impact on its descendants was sharply polarised. The impact was seen as substantial by some and as non-existent by others. The position that knowledge of development was irrelevant to the understanding of evolution was forcefully set out by the advocates of the Modern Synthesis, as we mentioned in Chapter 1. John Maynard Smith (1982) suggested that the widespread acceptance of August Weismann's (1885) doctrine of the separation of the germline from the soma was crucial to this line of thought. It led to the view that genetics, and hence evolution, could be understood without understanding development. These views were, until recently, dominant and involved the fusion of Darwin's mechanism for natural selection with Mendelian genetics. Briefly put, genes influence the characteristics of the individual; if individuals differ because of differences in their genes, some may be better able to survive and reproduce than others and, as a consequence, their genes are perpetuated.

The extreme alternative to the Modern Synthesis is a caricature of Lamarck's views about biological evolution and inheritance. If a blacksmith develops strong arms as a result of his work, his children will have stronger arms than would have been the case if their father had been an office worker. This view has been ridiculed by essentially all contemporary biologists. Nevertheless, as so often happens in polarised debates, the excluded middle ground concerning the evolutionary significance of development and plasticity has turned out to be much more interesting and potentially productive than either of the extreme alternatives. This view was developed at length by West-Eberhard (2003), was well-expressed in Gilbert and Epel's (2009) book, and is one that we extend here.

While the conventional model is that it is the evolution of the genotype that leads to phenotypic change, the possibility that phenotypic change might drive evolutionary processes and be reflected in genotypic change has increasingly become a point of focus (West-Eberhard, 2003, 2005a; Bateson, 2010). Given that plasticity is the basis of phenotypic change in the absence of genotypic change, such a model would give a central role to development in evolutionary processes.

In this chapter we consider these ideas and develop them further in the light of modern knowledge about epigenetics. The chapter begins with a discussion of the evolutionary significance of environmentally induced phenotypic plasticity. Epigenetic change induces phenotypic variation on which Darwinian selection can operate. We explore ways in which developmentally plastic responses and epigenetic change might become fixed in the genome.

CHOICE, CONTROL OF THE ENVIRONMENT AND NICHE CONSTRUCTION

Animals make active choices and the results of their choices can have consequences for subsequent evolution. The role of choice in evolution was clearly recognised by Darwin (1871) in his work on the principles of sexual selection. His thesis was that if members of one sex prefer to mate with members of the opposite sex that have a particular characteristic, such as having a longer tail than others, that characteristic will tend to become exaggerated in the course of evolution. Parenthetically, Alfred Russel Wallace, who arrived independently of Darwin at the principle of natural selection, never liked Darwin's ideas about sexual selection.

For many years, the process was regarded as insignificant, but most modern writers now accept sexual selection as an important evolutionary mechanism. Theoretical interest in sexual selection and empirical evidence for its importance has exploded in recent years (e.g. Andersson, 1994; Cornwallis and Uller, 2010). In sexual selection, differential survival of individuals is not at issue; rather the process depends on differential reproductive success. The extension of the selection terminology to such evolutionary processes has caused difficulties, and the things that animals do (mate choice and sexual conflict) need to be clearly distinguished from their consequences in terms of evolutionary change (sexual selection; see Bateson, 1983). A question remains about whether or not genetic changes underlying the characteristics of the chosen follow the actions of the choosers. Clearly sexual selection could not occur without variation in the characteristics of the chosen. However, the behaviour of the individuals choosing a mate determines which of the variants increases in frequency and it is therefore crucial to the changed allele frequency in the population.

This concept of choice and behaviour driving evolutionary change has been favoured and extended by some biologists over the years. The pioneering ecologist Charles Elton (1930) pointed out that the choice of environment by the animal was important in evolution. Such a sentiment was congenial to ethologists, who consistently placed emphasis on the active animal that engages with its environment. For example, the ethologist Robert Hinde (1959) drew attention to the variety of ways in which the choices of animals could influence the subsequent course of evolution. Waddington (1959) was also sympathetic to such a view, which emerged from his intense interest in forging links between developmental and evolutionary processes at a time when it was unfashionable to do so. He argued that animals choose the particular habitat in which their life will be passed. Thus the animal, by its behaviour, contributes in a most important way to determining what happens to the evolution of its own lineage.

More recently, interest has been drawn to the driving effect on evolution of a prey species making itself conspicuous to its predators. For instance, Fitzgibbon and Fanshaw (1988) pointed out that the curious leap of the Thompson gazelle when approached by a predator, known as 'stotting', indicates to a cheetah that it is less likely to catch that gazelle than one that does not jump or that does not jump as high. This gives those gazelles that make their jumps even more conspicuous an advantage over their less energetic peers. If each cheetah in each generation is able to learn that chasing high-jumping gazelles leads to a waste of effort and hunger, no evolutionary change is required in the cheetah and the change in the anti-predator behaviour of the gazelle is likely to evolve quickly. The driver is the choice by the predator of the less conspicuous prey. Similar evolutionary processes have been proposed for many other predator–prey interactions (e.g. Cresswell, 1994; Sherratt, 2002).

Many organisms change the physical or social conditions with which they and their descendants have to cope, thereby affecting the subsequent course of their evolution or that of other species. The environment does not simply set a problem to which the organism has to find a solution. The organism can do a great deal to create an environment to which it is best suited (Lewontin, 1983). By leaving an impact on their physical and social environment, organisms may affect the evolution of their own descendants, apart from changing the conditions for themselves. Sometimes the impact is subtle, such as when a plant sheds its leaves which, on falling to the ground, change the characteristics of the soil in which its own roots and those of its descendants grow. At other times the impact is conspicuous and active, such as when beavers dam a river, flood a valley and create a private lake for themselves. Modern beavers have webbed feet and it can be argued that this characteristic evolved in response to conditions created by their ancestors.

These ideas have been developed extensively over the past decade and are now referred to as 'niche construction' (Odling-Smee et al., 2003). Virtually all living organisms modify their immediate environments to some degree, and in many cases such impacts sum up across individuals to affect the evolution of their descendants and other populations inhabiting the same environment. Conventional evolutionary approaches encourage the view that such effects must be driven by genetic change. The niche construction perspective opens up a much broader range of explanations and suggests a mechanism whereby the organism creates a robust environment which is relatively immune to climatic and other variation that might otherwise impact on development.

The effect of controlling the environment on evolutionary change could be especially great when a major component of the environmental conditions with which animals have to cope is provided by their social environment. A similar type of directional response to that flowing from the effects of mate choice could operate in such circumstances (Jolly, 1966). If individuals compete with each other within a social group and the outcome of the competition depends in part on each individual's capacity to predict what the other will do, the evolutionary outcome might easily acquire a ratchet-like property. As Nicholas Humphrey (1976) noted, such an explanation makes sense of the astonishing rate of increase in the cranial capacity of humans, if it is assumed (reasonably) that cranial capacity and intellectual ability are correlated. Here again these ideas have been developed extensively (e.g. Byrne, 2000).

In a different context, many authors have emphasised how the cultural environment of humans has affected the human genome (see Laland et al., 2010). A much-cited example is the way in which herding cattle and drinking their milk influenced the evolution of the gene that is required for the synthesis in adult life of lactase, the enzyme required for digesting milk. In human populations that are not descended from cattle owners, such as the Chinese, the gene is not expressed in adulthood and these people are unable to digest cow's milk.

THE IMPORTANCE OF ADAPTABILITY

As is once again becoming obvious, climate can change dramatically with widespread and often disastrous consequences. The acidification of the oceans in the Permian (299–251 million years ago) led to a massive extinction in which 95% of existing marine animals died. Conditions that were fundamental to normal development changed and the effects on many organisms were lethal. The survivors may have had markedly different phenotypes from their immediate ancestors – as do those individuals exposed to other unusual conditions, such as toxins, early in development. As West-Eberhard (2005b) has emphasised, these survivors would have 'coped' with the inadequacies in their own characteristics.

Years earlier, James Mark Baldwin (1902) had called such individual responses 'accommodation'. If costs were involved in dealing with a challenge, and natural variation meant that some individuals coped at lower cost, then natural selection would have occurred and genetic changes would have taken place in the population. Costs in such cases might be energetic or the time required by the individual to acquire the necessary adaptations. That much is common ground for most biologists, though many have not considered the issue any further.

Organisms were doubtless usually passive in the initial stages of biological evolution driven by environmental change, but they could also have been active. This is the key conceptual point in understanding how plasticity and behaviour can drive evolutionary change. By their mobility, in the case of animals, or facility to disperse, in the case of plants (see, for example, Donohue, 2005), organisms often expose themselves to new conditions that may reveal heritable variability and open up possibilities for evolutionary changes that would not otherwise have taken place. For that reason, behaviour was likely to have been important in initiating evolutionary change in animals. Fish could move from a marine to a freshwater environment, herbivorous mammals could move into grasslands where the plants' defences involved toxins not previously experienced by the animals, and so forth. Our point is simply that the animals could be active agents in the evolutionary change of their descendants.

GENETIC CONSEQUENCES

A straightforward Darwinian account can be given of changes in the genome that follow a change in the phenotype driven by such environmental change. Consider the response of a population to an environment in which occasional fluctuations in conditions affect the mortality rate of juveniles to a much greater extent than that of adults. In such a situation it becomes advantageous for adults to release young into several different environments to maximise the chance that some will survive. The theory of bet-hedging addresses the issue of how individuals should optimise fitness in varying and unpredictable environments (Slatkin, 1974). A response enabling random switching between phenotypic states is an evolved strategy for coping with variable environments in some rapidly reproducing organisms such as bacteria. For example, the bacterium *Pseudomonas fluorescens*, when subjected to successive rounds of alternating selection in the presence and absence of shear stress, eventually gives rise to colonies that spontaneously switch between capsulated and non-capsulated morphologies (Beaumont et al., 2009). However, such 'bet-hedging' is an ineffective strategy when the organism has a high level of investment in offspring, is a slow reproducer, or where the environmental change has some level of predictability.

Much more controversial are the explanations given for adaptive changes made in response to environmental change when the adaptive modification of the phenotype, triggered by the environment in earlier generations is, in the course of evolution, seen in later generations. These ideas have a long history but are only now being integrated into mainstream evolutionary thought (Pigliucci and Murren, 2003). Key historical thinkers include Waddington and Schmalhausen. West-Eberhard, who has made important contributions herself (West-Eberhard, 2003), would also include James Mark Baldwin.

Waddington performed several now classic experiments investigating the impact of development on evolution. In one of them, he applied heat shock to the larvae of the fruit fly *Drosophila* (Waddington, 1953). This led to the development in some of the survivors of a peculiar character, namely the lack of a cross-vein in the adult wing. This character would not be considered an adaptation in the adult to larval heat shock; however, by selectively breeding these cross-veinless flies and therefore artificially ensuring the survival of the cross-veinless lineage, the character had a beneficial effect under his laboratory conditions. Waddington continued to apply heat shock in successive generations and to breed selectively from the flies with cross-veinless wings. After 14 generations, cross-veinless wings developed spontaneously in the absence of the external triggering condition of heat shock. Waddington called this process of change from environmental induction of the phenotype to its spontaneous expression, 'genetic assimilation'.

Waddington envisaged that, without larval shocking, the veins in the adult wing developed robustly with a consistent vein pattern – that is, the developmental process was well canalised. With larval heat shocking, the developmental process became less well canalised and a new phenotype was expressed in some of the descendant lineage. A possible mechanism is provided by studies using genetic manipulation of the heat shock protein Hsp90 in *Drosophila*, which suggest that lowered levels of that protein, the effects of which include altered histone modifications, are associated with greater phenotypic variation (Sollars et al., 2003). In time, the genes that were required for expressing the cross-veinless pattern were, it was suggested, brought into play. A conventional explanation in evolutionary biology might be that under intense artificial selection, the flies that developed the cross-veinless wings most readily were the ones that were most likely to survive the selection regime, and hence express the character robustly at a lower thermal threshold and eventually without the intervention of larval heat shock. Selection had, in effect, been on the genetic control of the thermal sensitivity necessary to express a hidden set of genomic determinants of the cross-vein phenotype. The alternative explanation advanced conceptually by Waddington, and then more mechanistically by later evo-devo advocates, was that an epigenetic change had been induced by the thermal shock, and upon artificial selection over generations, spontaneous genomic mutation occurred, thus *fixing* the epigenetic change as a genomic change. We shall provide a further possible explanation of fixation later in this chapter.

Giving credit to Baldwin (1902), West-Eberhard argued that after developmental disruption, the reorganisation of the genome might have a much broader effect than that envisaged by Waddington. She suggested that major changes might evolve as the character in question became more variable; in other words it became developmentally less robust or less canalised. The umbrella term that she used for all the heritable changes that might occur in the genetic regulation of development in response to environmental influences was 'genetic accommodation'.

One instance of the broadening effect suggested by West-Eberhard would be the evolution of a variety of distinct developmental trajectories or polyphenisms, each a canalised response and each expressed under specific conditions. A striking example of genetic accommodation was provided by Suzuki and Nijhout (2006), who studied tobacco hornworm caterpillars, the larvae of the moth *Manduca sexta*. These larvae are normally green, but a naturally occurring mutant form is black; both colour forms are heritable, with the black form arising from a sex-linked recessive allele that reduces juvenile hormone secretion, resulting in a black cuticle. When Suzuki and Nijhout heat-shocked the black larvae at the fourth larval instar, the sensitive period for the effect of juvenile hormone on cuticle coloration, some of the fifth larval instars were greenish. The authors bred from the greenest larvae, as well as those that did not show colour change on heat shock (that is, persistent black larvae), and monitored in each generation how the larvae appeared after heat shock and after being reared at different temperatures. By the 13th generation, the line selected for blackness developed robustly as black larvae irrespective of the rearing temperature. At high rearing temperatures, the line selected for greenness was more strikingly green than an unselected line that acted as a control group. However, at low rearing temperatures, the green-selected line was mostly black (although not as black as the black-selected line; Figure 8.1). Thus, the selection process enabled two different phenotypes to evolve at different constant temperatures. The green-to-black colour switch could be seen at a lower temperature, and within a smaller temperature range, compared with the control group.

Biochemical analyses revealed that the emergence of this polyphenism by genetic accommodation was due to a changed sensitivity to juvenile hormone. The genetic basis for this effect was investigated with backcross experiments, where frequency differences between female and male offspring colour suggested that the colour polyphenism was due to a sex-linked recessive gene of major effect; however, the presence of some intermediate-coloured offspring also indicated that genes of smaller effect exerted some influence over the phenotypic output of the major effect (Suzuki and Nijhout, 2008). The experiment was interpreted by the authors as showing that a spontaneous mutation in hormonal regulation, leading to the naturally occurring black morph, changed the hormonal pattern such that an environmentally induced polyphenism could be revealed. The process of artificial selection led to selection on modifier genes that changed the threshold effect and thus

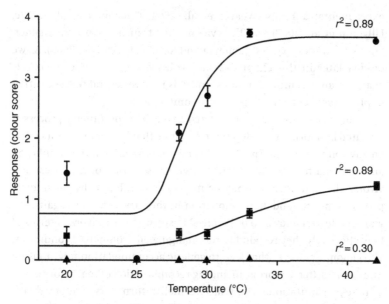

Figure 8.1. The reaction norms to temperature of 13th-generation green-selected (circles), black-selected (triangles) and control (squares) lines of the moth larvae (*Manduca sexta*). The green-selected and black-selected lines had been exposed to heat shock and artificially selected respectively for increased greenness or decreased greenness; the control group had not been subject artificial selection. From Suzuki and Nijhout (2006), reprinted with permission from the American Association for the Advancement of Science (AAAS). The original article did not mention the important detail that the tested larvae were not exposed to heat shock. Y. Suzuki (personal communication) has confirmed that this was indeed the case, showing that the observed phenotypes were inherited.

allowed the colour effect to occur at a lower thermal threshold. They argued that just as Hsp90 acts as a buffer to hide genetic variation, so too does the system regulating juvenile hormone secretion and action. In other words, the origin of polyphenisms may not depend on a new mutation but rather in subtle changes in regulatory systems.

However, a further interpretation, not considered by the authors, is that epigenetic mechanisms may be involved: either in inducing change in the regulatory pathways before fixation, or in sustaining the pathways throughout the study, as heat shock was applied in each generation. The action of Hsp90 appears to involve histone modifications (Sollars et al., 2003).

A wide variety of changes in endocrine regulation following developmental stresses are mediated by epigenetic mechanisms in experimental animals (Weaver et al., 2004; Gluckman et al., 2007c; Lillycrop et al., 2007) and the evidence for transmission across generations continues to grow (Jablonka and Raz, 2009; Nadeau, 2009). As we discuss later in this chapter, some evidence suggests that epigenetic changes can bias mutation rates, and this may be central to the fixation of phenotypic change via genetic accommodation.

In each of the cases considered here, the evolutionary process is initiated by a form of developmental change that is generally induced by an environmental cue. This leads to greater variability in the population and can lead to straightforward Darwinian selection of the fittest survivors. The new variants may or may not be followed by phenotypic plastic responses and accommodation in the survivors. Here again Darwinian selection may also be involved in sifting the survivors in terms of how effectively they respond to the changing environmental conditions. The genomic changes that lead to genetic accommodation might take a number of different forms, including recombination of regulatory genes or base-pair mutations. The issue we shall return to is whether this is simply traditional selection on pre-existing genetic variation, whether it is selection based on new genetic variation that is in turn based on a stochastic mutation, or it is the result of biases in mutation underpinned by pre-existing epigenetic change. However, before doing so, we shall consider a parallel set of phenomena involving behavioural adaptation.

ORGANIC SELECTION AND THE ADAPTABILITY DRIVER

Darwin supposed that if animals learned to perform an activity, generation after generation, then eventually the behaviour would be expressed without the necessity for learning. He did not explain how such a process might work. The first evolutionary explanation was proposed by Spalding (1873). His article is also historically important because it provides the first clear account of behavioural imprinting with which Konrad Lorenz (1935) is typically associated.

Spalding's driver of evolution comprised a sequence of learning followed by differential survival of those individuals that expressed the phenotype more efficiently without learning. The same idea was advanced once again by Baldwin (1896), Conwy Lloyd Morgan (1896) and Henry F. Osborn (1896), all publishing in the same year. Seemingly their ideas were advanced independently of Spalding and, indeed, of each other, although they may have unconsciously assimilated what Spalding wrote 23 years earlier in what was a widely read journal – *Macmillan's Magazine*, the predecessor of today's *Nature*.

Regardless of how they derived their ideas, the evolutionary mechanism proposed by Spalding and then Baldwin, Lloyd Morgan and Osborn was known at the time as 'organic selection' and is now frequently termed the 'Baldwin effect'. Lloyd Morgan's (1896) account of the process was particularly clear. He suggested that if a group of organisms respond adaptively to a change in environmental conditions, the modification will recur generation after generation in the changed conditions, but the modification will not be inherited. However, any variation in the ease of expression of the modified character which is due to genetic differences is liable to act in favour of those individuals that express the character most readily. As a consequence, an inherited predisposition to express the modifications in question will tend to evolve. The longer the evolutionary process continues, the more marked will be such a predisposition. Plastic modification within individuals might lead the process and a change in genes that influence the character would follow; one paves the way for the other.

An early example of how the process might work was provided by the child psychologist Jean Piaget, who began his career as a biologist and was much influenced by Baldwin. He studied the freshwater snail *Limnaea* and found that in still ponds the coiled shell of the form known as *stagnalis* was elongated. In lakes where wave action could render such a shell disadvantageous, the shell of the form known as *lacustris* was much shorter and the snail could cling onto the substrate much more firmly. When the *lacustris* form was reared in a still aquarium, half of the snails developed the shell of the *stagnalis* form, whereas the other half developed the more compact *lacustris* form without needing to be exposed to the wave action of a lake. One interpretation of this result is that the population of snails living in the lakes was mid-way through an evolutionary change. On this view the *stagnalis* form was evolutionarily more primitive, and in the lakes with wave action, some snails developed the *lacustris* adaptation as a result of developmental plasticity, but the evolutionarily more advanced snails no longer required such plasticity to develop their advantageous shell shape. Although Piaget carried out this work early in his career, it did not come out in book form until the end of his life (Piaget, 1979). In the terms used by Lloyd Morgan, the initial change could involve adaptability by the individual snail; the adaptability is won at some cost so that descendants expressing the character more efficiently would be more likely to survive.

Given Spalding's (1873) precedence and the simultaneous appearance in 1896 of the ideas about 'organic selection', it seems inappropriate to term the evolutionary process the 'Baldwin effect'. The trouble is that calling the proposed process the 'Spalding effect' is not descriptive of what initiates the hypothetical evolutionary process. 'Genetic accommodation', as discussed above, is a more general term which makes no inference about the inducing pathway; it would therefore be more appropriate to employ a term that captures the adaptability of the organism in the evolutionary process, and to this end, the term 'adaptability driver' has been suggested by Bateson (2006).

The zoologist Alister Hardy (1965) was an enthusiastic advocate of this process, as was the developmental psychobiologist Gilbert Gottlieb (1992). They envisaged a cascade of changes flowing from the initial behavioural event. Even without structural change, control of behavioural development might alter over time. A requirement for this to happen is that maintaining the plastic capacity to adapt to the new conditions would be more costly than doing it without plasticity. One instance might involve differential responsiveness to particular types of food. Many cases of the choice of an unusual food for a given species are probably not due to genetic changes, but to the functioning of normal mechanisms in unusual circumstances. A group of animals might be forced into living in an unusual place after losing their way, but they cope by changing their preferences to suitable foods that are locally abundant. Later, those descendants that did not need to learn so much when foraging might be more likely to survive than those that could only show a fully functional phenotype by learning. A cost would have been incurred in the time taken to learn. As a consequence, what began as a purely phenotypic difference between animals of the same species living in different habitats becomes a genotypic difference. Excellent examples of how a change in diet leads to the evolution of new digestive enzymes are given in Laland et al. (2010).

While the focus of Baldwin, as a psychologist, was largely on behaviour as the form of phenotypic response that was in some way, over time, incorporated into the genome, the model also allows for other forms of adaptive or plastic response to be thus incorporated. All that is required is that the adaptability in some way confers advantage in the novel environment, be it a physiological response such as coping with high altitudes by enhancing the oxygen-carrying capacity of the blood, or a change in coloration that improves concealment against predators, or a change in tail morphology in the tadpole that reduces the risk of predation. Over time, genetic accommodation can fix the alteration in the lineage. As the evolutionary change progressed, the population would consist of so-called 'phenocopies': individuals with the same phenotype but which developed in different ways.

CRITIQUE OF THE ORGANISM'S ROLE IN EVOLUTION

George Gaylord Simpson (1953), who forged the Modern Synthesis along with many others, did not think that development or behaviour played an important role in evolution. This became the standard line of neo-Darwinists. Simpson argued that if a new phenotype were valuable to the organism, it would evolve along well-established Darwinian lines. Secondly, he argued that if plasticity were a prerequisite for the evolutionary process and was generally beneficial, it would be disadvantageous to get rid of it.

On Simpson's first point, if learning (as an example of one form of adaptability in animals) involves several sub-processes or steps, as in operant chaining, then the chances of an unlearned equivalent appearing in one step in the course of evolution are small. However, with the learned phenotype as the standard, every small step that cuts out some of the plasticity with a simultaneous increase in efficiency is an improvement. As a hypothetical example of how the setting of an end-point might work, suppose that the ancestor of the Galapagos woodpecker finch (*Cactospiza pallida*), which pokes sharp sticks into holes containing insect larvae, did so by trial and error, and its modern form does so without much learning. In the first stage, a naïve variant of the ancestral finch, when in foraging mode, was more inclined than other birds to pick up sharp sticks. This habit spread in the population by Darwinian evolution because those behaving in this fashion obtained food more quickly. At this stage the birds still had to learn the second part of the sequence. The second step is that a naïve new variant, when in foraging mode, was more inclined to poke sharp sticks into holes. Again this second habit spread in the population by Darwinian evolution. The end-result is a finch that uses a tool without each individual having to learn how to do so. Simultaneous mutations increasing the probability of two quite distinct acts (picking up sticks and poking them into holes in the case of the woodpecker finch) would be unlikely. Learning makes it possible for them to occur at different times. Without learning, having one act but not the other has no value. Empirical evidence suggests that the woodpecker finch is halfway down the evolutionary road from fully learned to fully spontaneous (Tebbich et al., 2001).

A clear case of adaptability driving evolutionary change may be that of the house finch (*Carpdacus mexicanus*). In the middle of the twentieth century, the finch was introduced to eastern regions of the USA far from where it was originally found on the west coast. It was able to adapt to the new and extremely different climate and spread up into Canada. The finch also extended its western range north into Montana, where it has been extensively studied. After initial phenotypic adaptation, the house finch populations spontaneously expressed the physiological characteristics that best fitted them to their new habitats (Badyaev, 2009).

The question remains: under what circumstances will fixation of a previously plastic phenotype occur? The chances that all the mutations or genetic reorganisations necessary to give rise to genetic fixation would arise at the same time are small. In the natural world, if a phenotype expressed spontaneously without being learned is not as good as the learned one (in the sense that it is not acquired more quickly or at less cost), then nothing will happen and fixation will not occur. If the spontaneously expressed phenotype is better than the learned one, evolutionary change towards fixation is possible. If learning involves several sub-processes, as well as many opportunities for chaining (the discriminative stimulus for one action becoming the secondary reinforcer that can strengthen another), then the chances against a spontaneously expressed equivalent appearing in one step are small. However, with the learning available to fill in the gaps of a sequence, every small evolved step that cuts out the need for a plastic component while providing a simultaneous increase in efficiency is an improvement.

As far as learning is concerned, Simpson's second point of criticism was based on an inadequate understanding of how behaviour is changed and controlled. The answer to those who think that the proposed evolutionary change would lead to a generalised loss of the ability to learn is, quite simply, that it would not. Learning in complex organisms consists of a series of sub-processes (Heyes and Huber, 2000). If an array of feature detectors are linked directly to an array of executive mechanisms as well as indirectly through an intermediate layer, and all connections are plastic (Bateson and Horn, 1994), then a particular feature detector can become non-plastically linked to an executive system in the course of evolution without any more generalised loss of plasticity.

It remains to be seen whether these replies to Simpson's objections can be applied as cogently to other forms of phenotypic change, where the plastic response has been physiological or anatomical. Whereas behaviour is inherently plastic to some degree, anatomical change such as that seen in many insect polyphenisms is not. Physio-logical mechanisms either may or may not be. For example, when a queen bee is removed from the hive, some worker bees may re-activate their ovaries but they cannot change their anatomical phenotype fully to that of a queen (Tautz, 2008). When a plastic change involves a system that does not have parallel architecture with built-in redundancies, then the cost of losing it could outweigh the benefits of increasing the efficiency of response to an environ-mental challenge.

MODELS OF CHANGE INVOLVING LEARNING

Theoretical and modelling approaches have also been used to explore how phenotypic change, such as behaviour, can be incorporated and fixed into the genome. The hypothesis of the adaptability driver has been modelled, both analytically (e.g. Paenke et al., 2009) and by simu-lation (e.g. Suzuki and Arita, 2007). In the past, the evolutionary process was often taken to be a slow accretion by selection of spontaneously expressed phenotypic elements underpinned by stochastic mutation. Emphasis was placed on how particular behaviour patterns initially acquired by learning could be expressed spontaneously without learn-ing in the course of subsequent generations. Developments initiated by the work of Hinton and Nowlan (1987) have shifted the focus to the way in which learning can accelerate the rate at which challenges set by the environment can be met (e.g. Paenke et al., 2009). This aspect of evolu-tionary process draws attention to the advantages of learning in a changing environment (e.g. Borenstein et al., 2008). An ability to cope with complex environmental challenges by means of learning opens up ecological niches previously unavailable to the organism; a point elo-quently expressed by Avital and Jablonka (2000). This exposure to novel environments would inevitably lead to the subsequent evolution by classical Darwinian processes of morphological, physiological and bio-chemical adaptations to those niches. Many theoretical studies have indicated how crucial learning could be in such an evolutionary process (e.g. Beltman et al., 2004).

Learning would have become more and more important as pheno-types increased in complexity. A corollary is that if learning were not possible until a certain level of complexity had evolved, then a sharp increase in the rate of evolution would occur at that point. Where an environmental challenge involved greater processing capacity by the brain, this organ too would be expected to evolve with greater rapidity.

On the assumption that a bigger brain ensures greater learning capacity, the rate of evolution should correlate positively with relative
brain size. This expectation is given some support by a study suggesting
that the taxonomic groups evolving most rapidly have the biggest
brains relative to body size (Wyles et al., 1983). The expectation is also
supported by the correlation between brain size and behavioural innovation in primates (Reader and Laland, 2002) and between brain size and
the capacity to cope with novel environments in birds (Sol et al., 2005).
At the molecular level the massive growth of brain size and cognitive
capacity in the hominin line compared with chimpanzees has been
accompanied by an expansion in the range of brain enzymes involved
in epigenetic regulation, and particularly the editing of non-coding
RNAs (Paz-Yaacov et al., 2010).

Some theorists have argued that plasticity could dampen down
the rate of evolution (e.g. Robinson and Dukas, 1999; Price et al, 2003).
Their idea was that, with every individual in the population coping
plastically with an environmental challenge, natural selection would
have had no variation on which to act. In some cases this might well
have been true in the short run. However, if operating plastic mechanisms involved time and energy costs, then individuals that expressed
the adaptation spontaneously would readily invade the population and
the dampening effect would be lost.

THE POTENTIAL ROLE OF EPIGENETIC PROCESSES IN DRIVING EVOLUTIONARY CHANGE

Developments in molecular epigenetics have provided substantial
insight into the molecular basis of developmental integration (see
Chapter 5). Epigenetic changes can lead to fixed changes in gene
expression where the gene is suppressed in a cell lineage under all
circumstances, as is the case after cell differentiation. Alternatively,
epigenetic changes may affect chromatin structure with quite subtle
but potentially evolutionarily important effects; for example, they may
be limited to a single regulatory site in the promoter and change the
binding of transcriptional activators or repressors that are only active
in particular situations. Such epigenetic changes need not change basal
gene expression but only affect the degree of change in expression
in particular physiological contexts. These epigenetic effects can
be induced by environmental events acting during development, and
the subsequent phenotypic effects may be adaptive or non-adaptive for
the individual.

Furthermore a growing body of evidence suggests that after induction, phenotypes regulated by epigenetic mechanisms can be transmitted from one generation to the next (Gluckman et al., 2007b; Bossdorf et al., 2008; Jablonka and Raz, 2009), and while less stable than genomic variations, such heritable effects could play an important role in evolutionary processes. However, as this is a recently emergent area of research, empirical data are still scarce (although growing). Therefore, most of the discussion to date has been largely theoretical (Jablonka and Lamb, 1995; Price et al., 2003; Badyaev and Uller, 2009). These authors detail various pathways by which epigenetic change could affect evolutionary processes.

Because some forms of epigenetic marks can pass meiosis (Harper, 2005), the potential exists for selection on the phenotype induced by the epigenetic mark itself. Harper (2005) reviewed the growing evidence for epigenetic inheritance. It is clear that evolutionary processes could operate on non-genetic forms of inheritance (Helanterä and Uller, 2010). Epigenetic inheritance through microRNAs leading to phenotypic effects over several generations has been reported in at least two mouse models. The first model studied mice possessing a *Kit* paramutation (a heritable, meiotically stable epigenetic modification resulting from an interaction between alleles in a heterozygous parent) that results in a white-spotted phenotype. Injection of RNA from sperm of heterozygote mice into wild-type embryos led to the white-spotted phenotype in the offspring, which was in turn transmitted to their progeny (Rassoulzadegan et al., 2006). In the second model, mouse embryos were injected with a microRNA that targets an important regulator of cardiac growth. In adulthood these mice developed hypertrophy of the cardiac muscle, which was passed on to descendants through at least three generations without loss of effect (Wagner et al., 2008). Furthermore the microRNA was detected in the sperm of at least the first two generations, thus implicating sperm RNA as the likely means by which the pathology is inherited. The possible involvement of sperm is also supported by observations that transgenerational genetic effects on body weight and appetite can be passed through the mouse paternal germline for at least two generations (Yazbek et al., 2010).

Some evidence also suggests transmeiotic passage of methylation marks. In male rats exposed *in utero* to the endocrine disruptor vinclozolin during the sensitive period for testis sex differentiation and morphogenesis, lowered spermatogenic capacity and several adult-onset diseases were observed through four generations, and these were accompanied by altered DNA methylation patterns in the germline (Anway et al., 2005). Further analysis of these male offspring revealed that vinclozolin decreased methylation levels of two paternally imprinted genes, and increased that of three maternally imprinted genes (Stouder and Paoloni-Giacobino, 2010). However, while the imprinting defects were transgenerational, some methylation levels were normalised by the F3 generation. This might suggest that the effect was not due to true epigenetic inheritance; rather, the germ-cell line that contributed to the formation of the F2 generation was exposed to the F0 environment while the F1 generation were fetal. This can occur irrespective of whether the F1 generation is male or female because in both sexes the inherited genome is sequestered within germ-line cells at the early stages of embryonic life.

In these model systems, the environmental stimulus is only applied in one generation. In natural conditions, the environmental cues that induce epigenetic change may be recurrent and buttress what has happened in previous generations. This recurring effect might stabilise the phenotype until genetic accommodation and fixation have occurred.

The evolution of epigenetic inheritance may involve complex phenotypes, as epigenetic changes often occur in regulatory rather than coding regions of the genome and thus lead to quite widespread effects. Epigenetic processes may be involved in the experiments initiated by Belyaev (1969) and his colleagues. They carried out artificial selection experiments on the silver fox (*Vulpes vulpes*), a variant of the red fox. They selected, for breeding purposes, foxes that were least timid when a gloved hand was thrust at them and attempts were made by humans to handle them. Within two to three generations, foxes in the selected line were much tamer. In the fourth generation some of the cubs responded to humans by dog-like tail wagging. As the experiment proceeded, more and more dog-like behaviour appeared. In the sixth generation, some cubs eagerly sought contacts with humans – not only wagging their tails, but also whining, whimpering, and licking in a dog-like manner. The changes in standard coat colour pattern appeared in the eighth to tenth generation of the selected line. Piebald star spotting and brown mottling on the background of the standard silver–black colour were typical. Floppy ears and curly tails occurred in addition to changes in standard coat colour. In later generations, changes in the skeletal system began to appear, including shortened legs, tail, snout, upper jaw, and a widened skull (Trut et al., 2009).

The precise mechanisms of genetic regulation involved in these changes might involve artificial selection of genes that have a variety of effects, a phenomenon known as 'pleiotropy'. Alternatively, as we suggest here, the regulatory processes of development have been changed by the selection regime. Either way, the outcome of this remarkable long-term experiment is highly relevant to understanding the rate of domestication of dogs and the development of the great variety of breeds that are seen today.

CAN EPIGENETIC CHANGE LEAD TO FIXATION?

A central question in considering evolutionary change driven by the environment is whether the transmitted epigenetic markers could facilitate genomic change (see Johnson and Tricker, 2010). The answer is that in principle they could if they were present in the germline, if they increased the fitness of the individual carrying the markers and genomic reorganisation enabled some individuals to develop the same phenotype at lower cost. That much is exactly the same as has been proposed for the operation of the adaptability driver.

Epigenetic inheritance may serve to protect the well-adapted phenotypes within the population until spontaneous fixation occurs. Evidence for this is found in studies by True and Lindquist (2000) on yeast, in which two structural forms of a protein (Sup35) involved in regulating protein translation are spontaneously found; both forms have different functional significance and one form is favoured and spreads under conditions of stress.

However, an additional and important point is that the DNA sequences where epigenetic modifications have occurred may be more likely to mutate than other sites. The consequent mutations could then give rise to a range of phenotypes on which Darwinian selection could act. If epigenetic change could affect and bias mutation rates, such non-random mutation would facilitate fixation.

Methylated CpGs are mutational hotspots due to the established propensity of methylated and hydroxymethylated cytosine to undergo spontaneous chemical conversion into thymine and hydroxymethylated uracil, respectively (Pfeifer, 2006). As these are functional nucleobases, they are not recognised as damaged DNA and excised or corrected by DNA repair mechanisms. Thus the mutation becomes incorporated in subsequent DNA replications (Jones et al., 1992). DNA mapping shows fewer CpG sequences in the DNA than expected (Schorderet and Gartler, 1992), and CpG hypermutability has led to a decrease in frequency of amino acids coded by CpG dinucleotides – a trend that is only seen in organisms that possess DNA methyltransferase (Misawa et al., 2008). Indeed, comparison of the human and chimpanzee genomes has shown that 14% of the single amino acid changes are due to the biased instability of CpG sequences, which can be subject to methylation and thence to mutations (Misawa and Kikuno, 2009). The methylation of CpGs is a major contributing factor to mutation in *RB1*, a gene in which allelic inactivation leads to the developmental tumour, retinoblastoma (Mancini et al., 1997).

Further evidence in support of the hypothesis that epigenetic change can lead to mutation is found in the analysis of neutrally evolving strands of primate DNA. The evidence indicates that the phylogenetically 'younger' sequences have a higher CpG content than the 'older' sequences, due to the reduced opportunity for spontaneous mutation. Intriguingly, the CpG content is strongly correlated with a higher rate of neutral mutation at non-CpG sites (Walser et al., 2008), which suggests that CpGs play a role in influencing the mutation rate of DNA not containing CpG, perhaps by influencing the chromatin conformation surrounding the CpG and making it more accessible to other modifying processes. Furthermore, CpG content also appears to influence the *type of mutation* that occurs, with a higher ratio of transition-to-transversion mutations observed in parallel with the non-CpG mutation rate (Walser and Furano, 2010).

Although the rate and process of mutation is intrinsically dependent on the DNA sequence itself, the specific mechanisms by which this effect is exerted and the implications for genome maintenance remain to be established. These effects can only be of evolutionary significance if the epigenetic modification, and thus mutational bias, occurs in the germline itself. Even so, ample evidence indicates that the germline is subject to epigenetic modification.

While empirical evidence relevant to the impact of developmental plasticity and epigenetic processes on evolution remains to be discovered, the circumstantial evidence is growing. The conversion of epimutations into mutations has been observed in some human cancers, for example internal cancers involving the tumour suppressor genes *p53* and *p16* (Gonzalgo and Jones, 1997). The movement of transposable elements – a key component of genetic change – is also dependent on open chromatin structure as well as genomic and environmental stresses (Capy et al., 2000; Liu et al., 2009). The steps involved in the suggested evolutionary process involving biased mutations are shown in Figure 8.2.

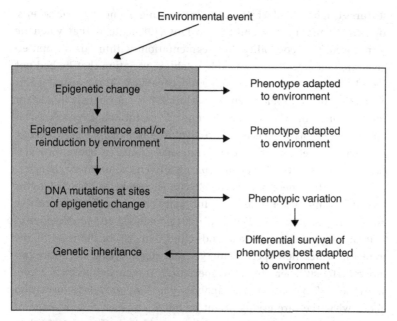

GENETIC/EPIGENETIC LEVEL   PHENOTYPIC LEVEL

Figure 8.2. Proposed steps by which an epigenetic change in response to
a challenge from the environment might eventually lead to genetic
inheritance. Phenotypic differences on which Darwinian selection
could act are rendered more probable by mutation rates being higher
at the sites of epigenetic change. The scheme does not propose, however,
that the phenotypes generated by genetic mutation would faithfully
copy the phenotypes generated by epigenetic change.

IMPLICATIONS FOR EVOLUTIONARY NOVELTY
AND SPECIATION

Major transitions in evolution have been explained in terms of
changes in genetic regulation early in development (Britten and Davidson, 1969) and offered as an explanation for the explosion of variety
seen in the Cambrian era (Carroll, 2005). Transitions in the rate of
evolution can involve the remodelling of existing structure by changes
in which part of a regulatory gene is expressed, and when in development it is expressed (Kirschner and Gerhart, 2005). Some of this might
involve epigenetic mechanisms (Zuckerkandl and Cavalli, 2007). The
occasional appearance of mutations and the reorganisation of the
genome permit evolutionary change that would not have previously
been possible. Gene duplication provides a substrate on which new

features can be added while sustaining existing phenotypic characteristics. In discussing evolvability, Dawkins (1989) noted that when he introduced the possibility for segmentation within his computer-generated biomorphs, he was able to obtain variation that he had not found without such a developmental capability. This general point about the role of development in evolution has enormously important implications for the understanding of evolutionary processes. It opened up the new science of 'evo-devo'. The role of epigenetic change in driving novel mutational substrates, as discussed above, provides further opportunities for phenotypically driven evolutionary change.

Speciation need not occur slowly. Phylogenetic branch lengths record the amount of time or evolutionary change between successive events of speciation. Venditti et al. (2010) analysed 101 phylogenies of animal, plant and fungal taxa and found that 78% of the trees fit the model in which new species emerge from single events, each rare but individually sufficient to cause speciation. This model predicts a constant rate of speciation. The analysis does not provide information about what the rare events might be but it emphasises how some of the drivers of evolution discussed here could be sufficient to lead to speciation.

Other studies show more speciation within a clade when polyphenism occurs within that clade (Pfennig and McGee, 2010). This suggests that the presence of developmentally induced polyphenism favours adaptive radiation, providing a range of niche-defined phenotypes on which Darwinian selection can act after fixation of the epigenetically mediated difference.

King (1993) suggested that speciation often involves a change in chromosome number. Closely related species can be strikingly different. In horses, for example, the chromosome number ranges from 32 in *Equus zebra hartmannae* and 46 in *Equus grevyi*, to 62 in *Equus assinus* and 66 in *Equus przewalski*; all but two of the horse hybrids are sterile. Similar variations in chromosomal number are also found in deer and mice (Duarte and Jorge, 1996; Kartavtseva and Pavlenko, 2000). Humans and chimpanzees have different chromosomal numbers – chromosome 2 of the human is a fusion of two ancestral chromosomes, denoted 2A and 2B in the chimpanzee (The Chimpanzee Sequencing and Analysis Consortium, 2005). How could these differences between closely related species arise in evolution? A hypothetical example illustrates one way.

Suppose that a herd of zebras wanders away from its usual habitat and enters an area where many of the plants available to the zebras as food contain toxins which they had not previously experienced.

These toxins exert a developmental impact on the fetuses carried by the mares and they form characteristics that are novel. When born, the zebra foals cope through phenotypic accommodation, but this nevertheless occurs at significant cost. In time, and in some individuals, these costs are minimised by genetic changes – perhaps biased by epigenetic change – and Darwinian selection operates to the advantage of these individuals and their offspring. Over time, the reorganisation required by such changes cascades and more and more genetic changes appear as the Darwinian process creates new order in the regulation of the zebra's development. The final step in this conjecture is that the genomic reorganisation impacts on chromosome number, which is under genetic control. If this happens, then a reproductive barrier would be established between the new zebra population and the one from which it originated.

Our general point, in which we follow Avital and Jablonka (2000), is that an individual's adaptability allows a lineage to occupy a new place which can then lead to descendants entering many unexploited niches within that new habitat. The Galapagos finches are a clear example of how, in a relatively short space of time, birds arriving from the mainland were able to radiate out into many different habitats (Grant, 1986). Tebbich et al. (2010) discuss how the finches' adaptability, for which they provide some evidence, could have played an important role in this process. The Darwinian triad of steps to explain the evolution of adaptations remains unchallenged. The postulated evolutionary process requires variation, differential reproductive success, and inheritance. However, the ways in which each step can come about have been revealed by advances in knowledge (see Pfennig et al., 2010).

CONCLUSIONS

One of the near-universal aspects of biology is the way in which genetically identical individuals are able to develop in such strikingly different ways. Phenotypic variation can be triggered during development in a variety of ways, some mediated through the parent's phenotype. Sometimes phenotypic variation arises because the environment triggers a developmental response that is appropriate to those conditions. Sometimes the organism 'makes the best of a bad job' in suboptimal conditions. Sometimes the buffering processes of development may not cope with what has been thrown at the organism, and a bizarre phenotype is generated. Whatever the adaptedness of the phenotype, each of these effects demonstrates how a given genotype will express itself differently in different environmental conditions.

The decoupling of development from evolutionary biology could not hold sway forever. Whole organisms survive and reproduce differentially and the winners drag their genotypes with them (West-Eberhard, 2003). As we have suggested, the way they respond phenotypically during development may influence how their descendants' genotypes evolved and were fixed. This is one of the important engines of Darwinian evolution and is the reason why it is so important to understand how whole organisms behave and develop.

We have explored how the characteristics of the organism affect the evolution of its descendants. The characteristics may be such that they constrain the course of subsequent evolution or they may facilitate a particular form of evolutionary change. The theories of biological evolution have been reinvigorated by the convergence of different disciplines. The combination of developmental and behavioural biology, ecology and evolutionary biology has shown how important the active roles of the organism are in the evolution of its descendants. The combination of molecular biology, palaeontology and evolutionary biology has shown how important an understanding of developmental biology is in explaining the constraints on variability and the direction of evolutionary change.

9

Development and evolution intertwined

One of the most remarkable characteristics of multicellular organisms is the way in which a single seed or a fertilised egg ends up as a fully functional complex organism, bigger in size by orders of magnitude. Science has been slow to uncover the processes involved in these remarkable transformations. As in all such inquiries, simplifications are necessary as aids to discovery. To modern eyes, some of these simplifications seem ridiculous. For example, the microscopists of the late seventeenth century convinced themselves that they could discern in the head of a human sperm a tiny homunculus – a miniature version of the adult. Everything was preformed and development was simply a matter of getting larger. The preformationists were opposed by those who believed in the emergence of an organism from an undifferentiated state. The tiny organism required interaction with the world around it if it was to develop successfully. This was known as 'epigenesis' – a term similar to modern 'epigenetics' which has, however, acquired a different meaning.

The next level of simplification is that the organism's characteristics require two sources of instruction: one from within and one from without. For most people, that is where understanding has not advanced. As noted at the beginning of this book, simplistic dichotomies abound: genetics or the environment, nature or nurture? The organisation of the brain is hard-wired or it is changeable. Development of the phenotype is robust or it is plastic. Some features of the adult organism require no input or 'commands' from the environment, while other features depend on such input. Behaviour is either innate or it is learned. It is well adapted or it is adaptable. In his excellent book about the changing role of the embryo in evolutionary thought, the philosopher Ron Amundson (2005) examines several more: genotype versus phenotype, germline versus soma, proximate versus ultimate, and typological thinking versus population thinking.

In the first chapter of this book we strongly criticised the conventional opposition of nature and nurture. These terms refer to different domains and should not be contrasted directly; one is a state and the other a process. Nature, we argued, should refer to the fully developed characteristics of an organism, and nurture to the ways in which those characteristics were derived. Even with this issue out of the way, the developmental processes are commonly separated into opposing categories. While we believe that it is unhelpful to separate complex phenomena into two classes, many distinguished scientists have thought otherwise.

In studies of human cognition, the philosopher and classicist Geoffrey Lloyd (2007) highlights the problems that arise when those who believe in human universals confront those who focus on human variation. The issues around which these debates revolve range from such matters as colour perception and spatial awareness to the classification of plants and animals. However, the debates that focus on the dichotomies of where adult characteristics come from are not about whether these characteristics are all of one category or the other but about whether or not the sharp distinctions drawn between the two categories are valid.

GENES AND THE ENVIRONMENT

Charles Darwin used the word 'innate' synonymously with 'inherited', and he used 'instinct' as synonymous with 'inherited behaviour' (Darwin, 1859, 1871, 1872). What does 'inherited' mean in this context? The view implicit in Darwin's writings is that some privileged material that determines the development of certain traits, including behavioural traits, is transferred from parents to offspring at the moment of conception. These traits are 'biologically inherited' or 'instinctive'. All other traits are 'acquired'. This was not just Darwin's view; it was the prevalent view in the nineteenth century (Mameli, 2005).

What Darwin famously and importantly added to conventional thinking was the proposal that inherited traits can evolve by a process of differential survival and reproduction. When the results of Mendel's breeding experiments were rediscovered and the science of genetics started in earnest, genes were identified with those properties that are transmitted from parents to offspring at conception and are exclusively responsible for the development of the inherited traits.

Konrad Lorenz, the great pioneering ethologist, drew a distinction between the adaptive information stored in the genome by the process of Darwinian evolution and adaptive information extracted from the environment. He wrote:

> No biologist in his right senses will forget that the blueprint contained in the genome requires innumerable environmental factors in order to be realized in the phenogeny of structures and functions. During his individual growth, the male stickleback may need water of sufficient oxygen content, copepods for food, light, detailed pictures on his retina, and millions of other conditions in order to enable him, as an adult, to respond selectively to the red belly of a rival. Whatever wonders phenogeny may perform, however, it cannot extract from these factors information which simply is not contained in them, namely, the information that a rival is red underneath. (Lorenz, 1965, p. 37)

In the different domain of human cognitive development, Noam Chomsky argued that it is not possible for children to extract information about 'permissible' syntactic rules from their linguistic environment. Therefore he concluded that syntactic knowledge must be derived from some internal source. Chomsky often mentions 'genetic determination' and 'genetic endowment' when writing about the innateness of the language faculty (e.g. Chomsky, 2000). If the information about syntactic structures is not extracted from linguistic stimuli, then it must be extracted from the genome. From the standpoint of modern biology, Chomsky's understanding of how genes are expressed was based on a vague notion of how genes influence developmental processes. Furthermore, Chomsky did not wish to explain the 'deep structure' of language in terms of Darwinian evolution. Nevertheless, the distinction that Chomsky sought to draw was clear enough, and was remarkably similar to the distinction drawn by Lorenz.

Why do we think that these great men were wrong about their distinctions between internal and external information? In part this is because 'environmental information' does not neatly decompose into that which carries critical inputs and that which is necessary merely to keep the organism going. In part it is because their distinctions encouraged the idea that the developmental processes leading to robust outcomes could be cleanly separated from those that were plastic.

PLASTICITY AND ROBUSTNESS

Many different forms of plasticity are now recognised, each dependent on markedly distinct types of experience. A robust phenotype is often supposed to be one produced by processes that are difficult to disrupt (developmental non-malleability), and one that is difficult to modify once it has developed (post-developmental non-malleability). However, as we have seen in Chapter 3, developmental and post-developmental robustness do not necessarily go together and are not based on unitary processes. A trait that is robust with respect to its development may not also be robust with respect to its continuance, and vice versa. As noted in Chapter 3, developmental malleability may be followed by non-malleability, as in many examples of alternative phenotypes found throughout the animal kingdom. The same is true for humans, as in the case of sexual differentiation or differentiation of the visual pathway. Conversely, developmental non-malleability may be followed by considerable malleability, as in the case of the human smile, which reliably appears in infants during the fifth or sixth week after birth and is successively greatly modified by social interactions and cultural influences (Bateson and Martin, 1999).

THE PROBLEM WITH METAPHORS

Our most fundamental reason for rejecting the distinction between internal and external sources of information lies at the heart of this book. We, like others (Pigliucci, 2010), reject the blueprint image because it implies a straightforward link between the starting points and the endpoints of development. Correspondences between plan and product are not to be found. In a blueprint, the mapping works both ways. Starting from a finished house, the room can be found on the blueprint, just as the room's position is determined by the blueprint. This straightforward mapping is not true for genes and behaviour, in either direction.

The language of a gene 'for' a particular trait, so often used by scientists, is exceedingly confusing to the non-scientist (and, if truth be told, to many scientists as well). What the scientists mean (or should mean) is that a genetic difference between two groups is associated with a difference in phenotype. They know perfectly well that other things are important and that the developmental outcome depends not only on the whole gene 'team' but also on the factors that remain the same from one experiment to the next or from one aeon to the next. Unfortunately, the language of genes 'for' characters has a way of seducing the scientists themselves into believing their own sound-bites.

The blueprint and programming images are misleading because they are too static and too suggestive of the idea that adult individuals are merely the inevitable expanded versions of fertilised eggs. In reality, individuals play an active role in their own development and, as now becomes obvious, in that of subsequent generations. Even when a particular gene or a particular experience is known to have a powerful effect on the development of a trait, biology has an uncanny way of finding alternative routes. Redundancy is commonly found in biology. If the normal developmental pathway to a particular form of adult behaviour is impassable, another way of getting to the same end-point is often found.

THE DYNAMICS OF DEVELOPMENT

At all levels of analysis, scientists studying development are faced with dynamic systems that alter their characteristics when conditions change. The way to understand such systems is by studying them as processes rather than by taking snapshots or by abstracting linear causal chains. One limitation of gene expression studies is that they are measures only of the molecular responses to the conditions at the time of sampling. For example, a gene that regulates red blood cell production is only recognised when the individual is short of oxygen. Otherwise it may not be expressed.

In the study of behavioural development, one approach, which now seems outmoded, was to make manipulations in early life and examine the long-term effects on adult behaviour. The conclusion emerging from such thinking, based on studies of populations, was that the genetic and environmental factors are *added together* to exert their effects on the behaviour of individuals. However, even if the dynamics of development could be plausibly reduced in this way, major aspects of the interplay between the developing individual and its environment were easily missed by the analytical techniques used to derive such conclusions.

The environment does not just provide the energy and matter required by developmental processes. It is also partly responsible for cueing which nuclear genes are switched on and off and the way in which the products of cellular transcription and translation are processed and used. This view was first demonstrated by Jacob and Monod (1961) in their work on genes controlling the production of enzymes needed to digest lactose by the bacteria *Escherichia coli*; these genes are repressed in the absence of lactose. More recent work in epigenetics has taken this insight much further. Environmental change can induce long-standing epigenetic change which, in turn, can affect how the organism responds to later environmental challenges. In rats subject to a high-fat diet after weaning, the degree of obesity that ensues is heavily influenced by the prenatal nutritional environment. The latter, in turn, influences the epigenetic state of specific genes, which leads to long-term effects on appetite and peripheral metabolism (Gluckman et al., 2009).

DEVELOPMENT IS COMPLEX

The opposite point of view to the simplifications we have criticised is extreme holism, in which it is supposed that everything interacts with everything else. This is a position we do not espouse. It is anti-analytical and does not help in understanding the mechanisms that interest us. Lest anybody should be in doubt, our views are not based on the assumption that all systems develop in the same way. Far from it. It would be hard to maintain, in the context of modern knowledge, that development can be reduced to just two different kinds of processes. To take two contrasting examples of plasticity: the triggering mechanisms involved in producing alternative phenotypes are unlikely to be similar to the many processes of learning. Indeed, learning mechanisms probably differ radically from each other. Furthermore, the processes involved in the development of some neurobiological systems are probably quite different from others that require plasticity and may be more akin to the robust development of an organ such as a kidney. Yet the outcome of development will often have involved interplay between the systems involved in robust development and those involved in plasticity. This conclusion does not lessen the value of identifying features that are important in development, such as robustness and plasticity, nor does it attempt to discover how they have evolved. It merely asserts that features may be co-opted for use in different ways and in different combinations.

The general point we wish to make may be helped by an analogy. Consider a rule-governed game such as chess. It is impossible to predict the course of a particular chess game from knowledge of the game's rules. Chess players are constrained by the rules and the positions of the pieces in the game, but they are also instrumental in generating the positions to which they must subsequently respond. Further it is a game played by two players with different degrees of ability to look ahead. The range of possible game sequences is enormous. The rules may be simple but the outcomes can be extremely complex. We have argued that order underlies even those developmental processes that make animals, including humans, different from each other.

Knowing something of the underlying regularities in development does bring an understanding of what happens to the child as it grows up. The ways in which learning is structured, for instance, affect how the child makes use of environmental contingencies and how the child classifies perceptual experience. Yet predicting precisely how an individual child will develop in the future from knowledge of the developmental rules for learning is no easier than predicting the moves in a chess game. The rules influence the course of a life, but they do not determine it. Like chess players, children are active agents. They influence their environment and are in turn affected by what they have done. Furthermore, children's responses to new conditions will, like chess players' responses, be refined or embellished as they gather experience. Sometimes the normal development of a particular ability requires input from the environment at a particular time; what happens next depends on the character of that input. The upshot is that, despite their underlying regularities, developmental processes seldom proceed in straight lines. Big changes in the environment may have no effect whatsoever, whereas some small changes can have big effects. The only way to unravel this is to understand the regulatory processes and their interplay with the processes that generate change.

Over-used metaphors from engineering such as 'hard-wiring' and 'pre-programming', applied globally to the outcome of development, fail to capture the character of the processes and once again invite the mistaken view that they can be contrasted with their opposites. A thorough investigation of developmental processes has been hindered by the indiscriminate use of two opposed classes of mechanisms, often typified by the 'is it nature or nurture?' question. The realisation that the many processes of development interact and are intertwined is crucial in order to make progress in biology and cognitive science. The dichotomies applied to phenotypic outcomes and the processes by which they were derived may or may not have played a positive role in helping to organise thinking in the past. In the current state of knowledge, the distinction hinders the scientific understanding of development. The use of the distinction generates in researchers the false illusion that certain important empirical questions have already been answered.

Much of the debate about development, even within the scientific community, has been dominated by 'folk psychology' and 'folk biology' (see Linquist et al., 2011). The conceptual flowerbed is full of vigorous

weeds with deep roots. Some energetic gardening is necessary. By bringing together evidence from different levels of analysis, ranging from the
molecular to the social, we have argued that a much more powerful
theoretical perspective can be formulated. Its impact is not only on
scientific approaches to development and evolution, but also on how
humans think about themselves and how they design public health
measures when mismatches between themselves and their environment occur. Robustness and plasticity are complementary and intertwined and must be considered together. They should not be seen as
being in opposition to each other.

As others, such as Keller (2010), have emphasised, population
arguments about sources of variation must be distinguished from
mechanistic issues dealing with how an individual develops. The concept of gene–environment interaction (G × E) arose from population
studies designed to uncover sources of variation. It is inappropriately
used for studies of developmental mechanisms in individuals. It is
particularly unfortunate when applied to cases where the organism
may change its environment (niche construction) or choose an environment to which it is best suited (niche picking). Genetic differences may
well be correlated with some phenotypic differences between humans,
but such knowledge is quite separate from knowledge of developmental
processes. Furthermore, regularities in development do not necessarily
imply regularities in the phenotype. As these issues are clarified, we
hope that many scientists who thought they disagreed with each other
will find that they do not.

BROAD PROBLEMS FOR HUMANITY

While this book has focused primarily on the role of numerous biological factors acting on individual development across the life course,
it would be remiss of us not to draw attention to the ways in which
these factors interact with those in the social environment to maintain
cycles of disadvantage in human populations. According to the World
Bank, 1.3 billion people are living in absolute poverty, having less than
US$1.25 per day on which to live (World Bank, 2009). Despite the efforts
of numerous well-intentioned organisations, the number of people
living in absolute poverty is not diminishing.

The problem is that unless all the sources of poverty are tackled in
concert, those people who are trapped in it will be unlikely to escape
from that trap (Dasgupta, 2009). The biological responses allowing early
survival in response to severe deprivation, as discussed in Chapter 4,

exacerbate the situation. The grandmother's nutritional and medical state (probably) and the mother's (certainly) will affect the child's characteristics. The necessary responses by the developing offspring in order to survive involve reduced investment in growth, reduced immune function, and reduced cognitive development. The individual is left at greater risk in multiple ways. The likelihood of developing infectious disease further inhibits the developing child's bodily and intellectual growth. Lack of adequate education will mean that such children do not acquire the skills that would help them to cope with the numerous challenges facing them. The chances of doing anything about their plight are slim.

The problems of poor nutrition, poor health, poor education and lack of opportunities combine and interact to leave an appalling humanitarian disaster. Even if it were possible to ignore the suffering that is generated, the wastage of human capital is enormous. This is an area where an understanding of the development of the individual needs to be brought together with developmental economics of human societies.

THE EVOLUTIONARY–DEVELOPMENTAL SYNTHESIS

In this book we have identified many ways in which robust outcomes can be generated. We have also identified many plastic processes. However, this analysis did not lead to a clear dichotomy of outcomes. Robust rules are required for plasticity to occur. This, in turn, may generate subsequent outcomes that are again highly robust. We considered the functional significance of the integrated mechanisms, and then – as a separate line of thought – how the component processes might have evolved. Finally, in what we regard as the most important part of integrating development and evolution, we examined the ways in which developmental mechanisms could be instrumental in driving evolutionary change.

In a much changed intellectual environment, the time seems right to rebuild an integrated approach to biology. With a whole array of promising new research areas and techniques emerging, integrative biologists have a lot to be excited about. This matters in a highly competitive world in which determined and well-placed people can, in a remarkably short time, change what is and what is not funded, open or close research institutes, and radically alter the departmental structure of universities. It is important, therefore, to offer to the new generation of young scientists who are coming into the field a sense of what is becoming once again one of the most exciting areas in biology.

With the enormous growth of knowledge about development, we find it appropriate to close this final chapter by adding two words to Theodosius Dobzhansky's (1973) famous dictum. It now becomes: 'Nothing in biology makes sense except in the light of evolution *and development*'.

References

Agrawal, A. A., Laforsch, C. & Tollrian, R. (1999). Transgenerational induction of defences in animals and plants. *Nature*, **401**, 60–63.

Albon, S. D., Clutton-Brock, T. H. & Guinness, F. E. (1987). Early development and population dynamics in red deer. II. Density-independent effects and cohort variation. *Journal of Animal Ecology*, **56**, 69–81.

Amundson, R. (2005). *The Changing Role of the Embryo in Evolutionary Thought: Roots of Evo-Devo*, Cambridge, UK: Cambridge University Press.

Andersson, M. (1994). *Sexual Selection*, Princeton, NJ: Princeton University Press.

Anway, M. D., Cupp, A. S., Uzumcu, M. & Skinner, M. K. (2005). Epigenetic transgenerational actions of endocrine disruptors and male fertility. *Science*, **308**, 1466–1469.

Applebaum, S. W. & Heifetz, Y. (1999). Density-dependent physiological phase in insects. *Annual Review of Entomology*, **44**, 317–341.

Arthur, W. (1997). *The Origin of Animal Body Plans: A Study in Evolutionary Developmental Biology*, Cambridge, UK: Cambridge University Press.

Avital, E. & Jablonka, E. (2000). *Animal Traditions: Behavioural Inheritance in Evolution*, Cambridge, UK: Cambridge University Press.

Badyaev, A. V. (2009). Evolutionary significance of phenotypic accommodation in novel environments: an empirical test of the Baldwin effect. *Philosophical Transactions of the Royal Society of London, Series B, Biological Sciences*, **364**, 1125–1141.

Badyaev, A. V. & Uller, T. (2009). Parental effects in ecology and evolution: mechanisms, processes and implications. *Philosophical Transactions of the Royal Society of London, Series B, Biological Sciences*, **364**, 1169–1177.

Baldwin, J. M. (1896). A new factor in evolution. *American Naturalist*, **30**, 441–451, 536–553.

Baldwin, J. M. (1902). *Development and Evolution*, London: MacMillan.

Barker, D. J. P. (1998). In utero programming of chronic disease. *Clinical Science*, **95**, 115–128.

Barker, D. J. P., Winter, P. D., Osmond, C., Margetts, B. & Simmonds, S. J. (1989). Weight in infancy and death from ischaemic heart disease. *Lancet*, **2**, 577–580.

Bateson, P. P. G. (1976). Rules and reciprocity in behavioural development. In *Growing Points in Ethology*, ed. P. P. G. Bateson & R. A. Hinde. Cambridge, UK: Cambridge University Press, 401–421.

Bateson, P. (1983). Optimal outbreeding. In *Mate Choice*, ed. P. Bateson. Cambridge, UK: Cambridge University Press, 257–277.

Bateson, P. (1987). Biological approaches to the study of behavioral development. *International Journal of Behavioral Development*, **10**, 1–22.

Bateson, P. (1994). The dynamics of parent–offspring relationships in mammals. *Trends in Ecology & Evolution*, **9**, 399–403.

Bateson, P. (2000). Models of memory: the case of imprinting. In *Brain, Perception, Memory: Advances in Cognitive Neuroscience*, ed. J. Bolhuis. Oxford: Oxford University Press, 267–278.

Bateson, P. (2001). Fetal experience and good adult design. *International Journal of Epidemiology*, **30**, 928–934.

Bateson, P. (2006). The adaptability driver: links between behaviour and evolution. *Biological Theory*, **1**, 342–345.

Bateson, P. (2007). Developmental plasticity and evolutionary biology. *Journal of Nutrition*, **137**, 1060–1062.

Bateson, P. (2010). The evolution of evolutionary theory. *European Review*, **18**, 287–296.

Bateson, P., Barker, D., Clutton-Brock, T., et al. (2004). Developmental plasticity and human health. *Nature*, **430**, 419–421.

Bateson, P. & Horn, G. (1994). Imprinting and recognition memory: a neural net model. *Animal Behaviour*, **48**, 695–715.

Bateson, P. & Mameli, M. (2007). The innate and the acquired: useful clusters or a residual distinction from folk biology? *Developmental Psychobiology*, **49**, 818–831.

Bateson, P. & Martin, P. (1999). *Design For a Life: How Behaviour Develops*, London: Cape.

Bateson, W. (1894). *Materials for the Study of Variation: Treated With Especial Regard to Discontinuity in The Origin of Species*, London: Macmillan.

Bauerfeind, S. S., Perlick, J. E. C. & Fischer, K. (2009). Disentangling environmental effects on adult life span in a butterfly across the metamorphic boundary. *Experimental Gerontology*, **44**, 805–811.

Beall, C. M. (2007). Two routes to functional adaptation: Tibetan and Andean high-altitude natives. *Proceedings of the National Academy of Sciences of the United States of America*, **104**, 8655–8660.

Beall, C. M., Cavalleri, G. L., Deng, L., et al. (2010). Natural selection on *EPAS1* (*HIF2α*) associated with low hemoglobin concentration in Tibetan highlanders. *Proceedings of the National Academy of Sciences of the USA*, **107**, 11459–11464.

Beaumont, H. J. E., Gallie, J., Kost, C., Ferguson, G. C. & Rainey, P. B. (2009). Experimental evolution of bet hedging. *Nature*, **462**, 90–93.

Belsky, J., Steinberg, L. & Draper, P. (1991). Childhood experience, interpersonal development, and reproductive strategy: an evolutionary theory of socialization. *Child Development*, **62**, 647–670.

Beltman, J. B., Haccou, P. & Ten Cate, C. (2004). Learning and colonization of new niches: a first step towards speciation. *Evolution*, **58**, 35–46.

Belyaev, D. K. (1969). Domestication of animals. *Science Journal (UK)*, **5**, 47–52.

Bird, C. D. & Emery, N. J. (2009). Rooks use stones to raise the water level to reach a floating worm. *Current Biology*, **19**, 1410–1414.

Blumberg, M. S. (2005). *Basic Instinct: The Genesis of Behavior*. New York: Thunder's Mouth Press.

Blumberg, M. S. (2009). *Freaks of Nature: What Anomalies Tell Us About Development and Evolution*, New York: Oxford University Press.

Bolhuis, J. J. (1991). Mechanisms of avian imprinting: a review. *Biological Reviews*, **66**, 303–345.

Borenstein, E., Feldman, M. W. & Aoki, K. (2008). Evolution of learning in fluctuating environments: when selection favors both social and exploratory individual learning. *Evolution*, 62, 586–602.

Bossdorf, O., Richards, C. L. & Pigliucci, M. (2008). Epigenetics for ecologists. *Ecology Letters*, 11, 106–115.

Breland, K. & Breland, M. (1966). *Animal Behavior*, New York: Macmillan.

Britten, R. J. & Davidson, E. H. (1969). Gene regulation for higher cells: a theory. *Science*, 165, 349–357.

Brönmark, C., Pettersson, L. B. & Nilson, P. A. (1999). Predator-induced defense in crucian carp. In *The Ecology and Evolution of Inducible Defenses*, ed. R. Tollrian & C. D. Harvell. Princeton, NJ: Princeton University Press, 203–217.

Brooks, A. A., Johnson, M. R., Steer, P. J., Pawson, M. E. & Abdalla, H. I. (1995). Birth weight: nature or nurture? *Early Human Development*, 42, 29–35.

Brown, R. W., Chapman, K. E., Edwards, C. R. & Seckl, J. R. (1993). Human placental 11-beta-hydroxysteroid dehydrogenase: evidence for and partial purification of a distinct NAD-dependent isoform. *Endocrinology*, 132, 2614–21.

Burkhardt, R. W. (2005). *Patterns of Behavior*, Chicago: University of Chicago Press.

Byrne, R. W. (2000). Evolution of primate cognition. *Cognitive Science*, 24, 543–570.

Campos, E. (1995). Amblyopia. *Survey of Ophthalmology*, 40, 23–39.

Cannon, W. B. (1929). Organization for physiological homeostasis. *Physiological Reviews*, 9, 399–431.

Capy, P., Gasperi, G., Biemont, C. & Bazin, C. (2000). Stress and transposable elements: co-evolution or useful parasites? *Heredity*, 85, 101–106.

Carroll, S. B. (2005). *Endless Forms Most Beautiful: The New Science of Evo Devo*, New York: Norton.

Cedar, H. & Bergman, Y. (2009). Linking DNA methylation and histone modification: patterns and paradigms. *Nature Reviews Genetics*, 10, 295–304.

Centerwall, S. A. & Centerwall, W. R. (2000). The discovery of phenylketonuria: the story of a young couple, two affected children, and a scientist. *Pediatrics*, 105, 89–103.

Chali, D., Enquselassie, F. & Gesese, M. (1998). A case-control study on determinants of rickets. *Ethiopian Medical Journal*, 36, 227–234.

Champagne, F. A. (2010). Epigenetic influence of social experiences across the lifespan. *Developmental Psychobiology*, 52, 299–311.

Champagne, F. A. & Curley, J. P. (2009). Epigenetic mechanisms mediating the long-term effects of maternal care on development. *Neuroscience and Biobehavioral Reviews*, 33, 593–600.

Champagne, F. A., Francis, D. D., Mar, A. & Meaney, M. J. (2003). Variations in maternal care in the rat as a mediating influence for the effects of environment on development. *Physiology & Behavior*, 79, 359–371.

Chomsky, N. (2000). *On Nature and Language*, Cambridge, UK: Cambridge University Press.

Chou, H.-H., Hayakawa, T., Diaz, S., et al. (2002). Inactivation of CMP-N-acetyl-neuraminic acid hydroxylase occurred prior to brain expansion during human evolution. *Proceedings of the National Academy of Sciences of the USA*, 99, 11736–11741.

Clutton-Brock, T. H., Guinness, F. E. & Albon, S. D. (1982). *Red Deer: Behaviour and Ecology of Two Sexes*, Chicago: University of Chicago Press.

Cohen, L. G., Celnik, P., Pascual-Leone, A., et al. (1997). Functional relevance of cross-modal plasticity in blind humans. *Nature*, 389, 180–183.

Coon, D. (2006). *Psychology: a Modular Approach to Mind and Behavior*, Belmont: Thomson Learning.

Cornwallis, C. K. & Uller, T. (2010). Towards an evolutionary ecology of sexual traits. *Trends in Ecology & Evolution*, **25**, 145–152.

Cresswell, W. (1994). Song as a pursuit-deterrent signal, and its occurrence relative to other anti-predation behaviors of skylark (*Alauda arvensis*) on attack by merlins (*Falco columbarius*). *Behavioral Ecology & Sociobiology*, **34**, 217–223.

Danial, N. N. & Korsmeyer, S. J. (2004). Cell death: critical control points. *Cell*, **116**, 205–19.

Darwin, C. (1859). *On the Origin of Species By Means of Natural Selection*, London: Macmillan.

Darwin, C. (1871). *The Descent of Man, and Selection in Relation to Sex*, London: Murray.

Darwin, C. (1872). *The Expression of the Emotions in Man and Animals*, London: John Murray.

Dasgupta, P. (2009). Poverty traps: exploring the complexity of causation. In *The Poorest and the Hungry: Assessments, Analyses, and Actions*, ed. J. von Braun, R. Vargas Hill & R. Pandya-Lorch, Washington, DC: International Food Policy Research Institute, 129–146.

Davenport, J. (1997). Temperature and the life-history strategies of sea turtles. *Journal of Thermal Biology*, **22**, 479–488.

Davis, O. S. P., Haworth, C. M. A. & Plomin, R. (2009). Dramatic increase in heritability of cognitive development from early to middle childhood: an 8-year longitudinal study of 8,700 pairs of twins. *Psychological Science*, **20**, 1301–1308.

Dawkins, R. (1976). *The Selfish Gene*, New York: Oxford University Press.

Dawkins, R. (1986). *The Blind Watchmaker*, Harlow: Longman.

Dawkins, R. (1989). The evolution of evolvability. In *Artificial Life VI: Proceedings of the Santa Fe Institute Studies in the Sciences of Complexity*, ed. C. Langton. Reading, MA: Addison-Wesley, 201–220.

De Kort, S. R., Tebbich, J. M., Dally, N. J., Emery, N. J. & Clayton, N. S. (2006). The comparative cognition of caching. In *Comparative Cognition: Experimental Explorations of Animal Intelligence*, ed. E. A. Wasserman & T. R. Zentall. Oxford: Oxford University Press, 602–618.

Dejong, G. (1995). Phenotypic plasticity as a product of selection in a variable environment. *American Naturalist*, **145**, 493–512.

Denver, R., Mirhadi, N. & Phillips, M. (1998). Adaptive plasticity in amphibian metamorphosis: response of *Scaphiopus hammondii* tadpoles to habitat desiccation. *Ecology*, **79**, 1859–1872.

Diamond, J. (1991). Pearl Harbor and the Emperor's physiologists. *Natural History*, **1991**, 2–7.

Dirani, M., Tong, L., Gazzard, G., et al. (2009). Outdoor activity and myopia in Singapore teenage children. *British Journal of Ophthalmology*, **93**, 997–1000.

Dobzhansky, T. (1973). Nothing in biology makes sense except in the light of evolution. *American Biology Teacher*, **35**, 125–129.

Donohue, K. (2005). Niche construction through phonological plasticity: life history dynamics and ecological consequences. *New Phytology*, **166**, 83–92.

Dover, G. (1986). Molecular drive in multigene families: how biological novelties arise, spread and are assimilated. *Trends in Genetics*, **2**, 159–165.

Duarte, J. M. B. & Jorge, W. (1996). Chromosomal polymorphism in several populations of deer (genus *Mazama*) from Brazil. *Archivos de Zootecnia*, **45**, 281–287.

Edelman, G. M. (1987). *Neural Darwinism*, New York: Basic Books.

Eichler, E. E. (2001). Segmental duplications: what's missing, misassigned, and misassembled – and should we care? *Genome Research*, **11**, 653–6.

Eldar, A., Shilo, B.-Z. & Barkai, N. (2006). Elucidating mechanisms underlying robustness of morphogen gradients. *Current Opinion in Genetics & Development*, **14**, 435–439.

Elton, C. (1930). *Animal Ecology and Evolution*, Oxford: Oxford University Press.

Erhuma, A., Bellinger, L., Langley-Evans, S. C. & Bennett, A. J. (2007). Prenatal exposure to undernutrition and programming of responses to high-fat feeding in the rat. *British Journal of Nutrition*, **98**, 517–524.

Eriksson, J. G., Forsen, T. J., Osmond, C. & Barker, D. J. P. (2003). Pathways of infant and childhood growth that lead to type 2 diabetes. *Diabetes Care*, **26**, 3006–3010.

Felsenfeld, G. & Groudine, M. (2003). Controlling the double helix. *Nature*, **421**, 448–453.

Feng, S., Cokus, S. J., Zhang, X., et al. (2010). Conservation and divergence of methylation patterning in plants and animals. *Proceedings of the National Academy of Sciences of the USA*, **107**, 8689–8694.

Fernald, R. D. (2000). Evolution of eyes. *Current Opinion in Neurobiology*, **10**, 444–450.

Feuillet, L., Dufour, H. & Pelletier, J. (2007). Brain of a white-collar worker. *Lancet*, **370**, 262.

Fitzgibbon, C. D. & Fanshawe, J. H. (1988). Stotting in Thomson gazelles: an honest signal of condition. *Behavioral Ecology & Sociobiology*, **23**, 69–74.

Flatt, T. (2005). The evolutionary genetics of canalization. *Quarterly Review of Biology*, **80**, 287–316.

Fraga, M. F., Ballestar, E., Paz, M. F., et al. (2005). Epigenetic differences arise during the lifetime of monozygotic twins. *Proceedings of the National Academy of Sciences of the USA*, **102**, 10604–10609.

Frankel, N., Davis, G. K., Vargas, D., Wang, S., Payre, F. & Stern, D. L. (2010). Phenotypic robustness conferred by apparently redundant transcriptional enhancers. *Nature*, **466**, 490–493.

Freeman, T. C. B., Sengpiel, F. & Blakemore, C. (1996). Development of binocular interactions in the primary visual cortex of anaesthetized kittens. *Journal of Physiology*, **494P**, P18–P19.

Fuglsang, J. & Ovesen, P. (2006). Aspects of placental growth hormone physiology. *Growth Hormone & IGF Research*, **16**, 67–85.

Gale, C. R., Jiang, B., Robinson, S. M., Godfrey, K. M., Law, C. M. & Martyn, C. N. (2006). Maternal diet during pregnancy and carotid intima-media thickness in children. *Arteriosclerosis, Thrombosis, and Vascular Biology*, **26**, 1877–1882.

Gangaraju, V. K., Yin, H., Weiner, M. M., Wang, J., Huang, X. A. & Lin, H. (2010). *Drosophila* Piwi functions in Hsp90-mediated suppression of phenotypic variation. *Nature Genetics*, **43**, 153–158.

Gilbert, S. F. (2005). Mechanisms for the environmental regulation of gene expression: ecological aspects of animal development. *Journal of Biosciences*, **30**, 65–74.

Gilbert, S. F. & Epel, D. (2009). *Ecological Developmental Biology: Integrating Epigenetics, Medicine and Evolution*, Sunderland, MA: Sinauer Associates.

Gilbert, S. F., Opitz, J. M. & Raff, R. A. (1996). Resynthesizing evolutionary and developmental biology. *Developmental Biology*, **173**, 357–372.

Ginsburg, S. & Jablonka, E. (2007). The transition to experiencing. II. The evolution of associative learning based on feelings. *Biological Theory*, **2**, 231–243.

Giraldeau, L.-A., Valone, T. J. & Templeton, J. J. (2002). Potential disadvantages of using socially acquired information. *Philosophical Transactions of the Royal Society of London, Series B, Biological Sciences*, **357**, 1559–1566.

Gluckman, P. D. & Hanson, M. A. (2004). Maternal constraint of fetal growth and its consequences. *Seminars in Fetal & Neonatal Medicine*, **9**, 419–425.

Gluckman, P. D. & Hanson, M. A. (2005). *The Fetal Matrix: Evolution, Development, and Disease*, Cambridge, UK: Cambridge University Press.

Gluckman, P. D. & Hanson, M. A. (eds) (2006). *Developmental Origins of Health and Disease*, Cambridge, UK: Cambridge University Press.

Gluckman, P. D. & Hanson, M. A. (2010). The plastic human. *Infant and Child Development*, **19**, 21–26.

Gluckman, P. D., Hanson, M. A. & Beedle, A. S. (2007a). Early life events and their consequences for later disease: a life history and evolutionary perspective. *American Journal of Human Biology*, **19**, 1–19.

Gluckman, P. D., Hanson, M. A. & Beedle, A. S. (2007b). Non-genomic transgenerational inheritance of disease risk. *Bioessays*, **29**, 145–154.

Gluckman, P. D., Hanson, M. A. & Buklijas, T. (2010). A conceptual framework for the developmental origins of health and disease. *Journal of Developmental Origins of Health and Disease*, **1**, 6–18.

Gluckman, P. D., Hanson, M. A., Buklijas, T., Low, F. M. & Beedle, A. S. (2009). Epigenetic mechanisms that underpin metabolic and cardiovascular diseases. *Nature Reviews Endocrinology*, **5**, 401–408.

Gluckman, P. D., Hanson, M. A. & Spencer, H. G. (2005a). Predictive adaptive responses and human evolution. *Trends in Ecology & Evolution*, **20**, 527–533.

Gluckman, P. D., Hanson, M. A., Spencer, H. G. & Bateson, P. (2005b). Environmental influences during development and their later consequences for health and disease: implications for the interpretation of empirical studies. *Proceedings of the Royal Society of London. Series B, Biological Sciences*, **272**, 671–677.

Gluckman, P. D., Lillycrop, K. A., Vickers, M. H., et al. (2007c). Metabolic plasticity during mammalian development is directionally dependent on early nutritional status. *Proceedings of the National Academy of Sciences of the USA*, **104**, 12796–12800.

Godfrey, K. M., Gluckman, P. D. & Hanson, M. A. (2010). Developmental origins of metabolic disease: life course and intergenerational perspectives. *Trends in Endocrinology & Metabolism*, **21**, 199–205.

Godfrey, K. M., Gluckman, P. D., Lillycrop, K. A., et al. (2009). Epigenetic marks at birth predict childhood body composition at age 9 years. *Journal of Developmental Origins of Health and Disease*, **1**, S44.

Gollin, E. S. (ed.) (1981). *Developmental Plasticity: Behavioral and Biological Aspects of Variations in Development*, New York: Academic Press.

Gonzalgo, M. L. & Jones, P. A. (1997). Mutagenic and epigenetic effects of DNA methylation. *Mutation Research*, **386**, 107–118.

Goodier, J. L. & Kazazian, H. H., Jr. (2008). Retrotransposons revisited: the restraint and rehabilitation of parasites. *Cell*, **135**, 23–35.

Gottlieb, G. (1971). *Development of Species Identification in Birds*, Chicago: University of Chicago Press.

Gottlieb, G. (1992). *Individual Development and Evolution*, New York: Oxford University Press.

Gould, S. J. & Lewontin, R. C. (1979). The spandrels of San Marco and the Panglossian paradigm: a critique of the adaptationist programme. *Proceedings of the Royal Society of London, Series B, Biological Sciences*, **205**, 581–598.

Gould, S. J. & Vrba, E. (1982). Exaptation: a missing term in the science of form. *Palaeobiology*, **8**, 4–15.

Grandjean, V., Hauck, Y., Beloin, C., Le Hégarat, F. & Hirschbein, L. (1998). Chromosomal inactivation of *Bacillus subtilis* exfusants: a prokaryotic model of epigenetic regulation. *Biological Chemistry*, **379**, 553–557.

Grant, P. R. (1986). *Ecology and Evolution of Darwin's Finches*, Princeton, NJ: Princeton University Press.

Green, R. E., Krause, J., Briggs, A. W., Maricic, T., et al. (2010). A draft sequence of the Neandertal genome. *Science*, 328, 710–722.

Griffiths, J. S. & Mahler, H. R. (1969). DNA ticketing theory of memory. *Nature*, 223, 580–582.

Griffiths, P. E. & Stotz, K. (2006). Genes in the post-genomic era. *Theoretical Medicine and Bioethics*, 27, 499–521.

Grohmann, J. (1939). Modifikation oder Funktionsreifung? Ein Beitrag zur Klärung der wechselseitigen Beziehungen zwischen Instinkthandlung und Erfahrung. *Zeitschrift für Tierpsychologie*, 2, 132–144.

Guan, J.-S., Haggarty, S. J., Giacometti, E., et al. (2009). HDAC2 negatively regulates memory formation and synaptic plasticity. *Nature*, 459, 55–60.

Guerrero-Preston, R., Goldman, L. R., Brebi-Mieville, P., et al. (2010). Global DNA hypomethylation is associated with *in utero* exposure to cotinine and perfluorinated alkyl compounds. *Epigenetics*, 5, 539–546.

Gunn, T. R. & Gluckman, P. D. (1989). The endocrine control of the onset of thermogenesis at birth. *Baillière's Clinical Endocrinology and Metabolism*, 3, 869–886.

Gwinner, E. (1996). Circadian and circannual programmes in avian migration. *Journal of Experimental Biology*, 199, 39–48.

Hamilton, W. D. (1964). The genetical evolution of social behaviour. *Journal of Theoretical Biology*, 7, 1–16.

Hardy, A. (1965). *The Living Stream*, London: Collins.

Harper, L. V. (2005). Epigenetic inheritance and the intergenerational transfer of experience. *Psychological Bulletin*, 131, 340–360.

Heijmans, B. T., Tobi, E. W., Stein, A. D., Putter, H., et al. (2008). Persistent epigenetic differences associated with prenatal exposure to famine in humans. *Proceedings of the National Academy of Sciences of the USA*, 105, 17046–17049.

Helanterä, H. & Uller, T. (2010). The price equation and extended inheritance. *Philosophy & Theory in Biology*, 2, e101.

Helgason, A., Palsson, S., Guobjartsson, D., Kristjansson, P. & Stefansson, K. (2008). An association between the kinship and fertility of human couples. *Science*, 319, 813–816.

Hensch, T. K. (2004). Critical period regulation. *Annual Review of Neuroscience*, 27, 549–579.

Heyes, C. & Huber, L. (2000). *The Evolution of Cognition*, Cambridge, MA: MIT Press.

Hinde, R. A. (1959). Behaviour and speciation in birds and lower vertebrates. *Biological Reviews*, 34, 85–128.

Hinton, G. E. & Nowlan, S. J. (1987). How learning can guide evolution. *Complex Systems*, 1, 495–502.

Hölldobler, B. & Wilson, E. O. (1990). *The Ants*, Cambridge, MA: Harvard University Press.

Holmes, F. L. (1963). Claude Bernard and the milieu intérieur. *Archives Internationales d'Histoire des Sciences*, 16, 369–376.

Horn, G. & McCabe, B. J. (1984). Predispositions and preferences: effects on imprinting of lesions to the chick brain. *Animal Behaviour*, 32, 288–292.

Horn, G., Rose, S. P. R. & Bateson, P. P. G. (1973). Experience and plasticity in the central nervous system. *Science*, 181, 506–514.

Huang, S. (2009). Reprogramming cell fates: reconciling rarity with robustness. *BioEssays*, 31, 546–560.

Humphrey, N. K. (1976). The social function of intellect. In *Growing Points in Ethology*, ed. P. P. G. Bateson & R. A. Hinde. New York: Cambridge University Press, 303–317.

Jablonka, E. & Lamb, M. J. (1995). *Epigenetic Inheritance and Evolution: The Lamarckian Dimension*, New York: Oxford University Press.

Jablonka, E. & Lamb, M. J. (2005). *Evolution in Four Dimensions*, Cambridge, MA: MIT Press.

Jablonka, E., Oborny, B., Molnar, I., Kisdi, E., Hofbauer, J. & Czaran, T. (1995). The adaptive advantage of phenotypic memory in changing environments. *Philosophical Transactions of the Royal Society of London, Series B, Biological Sciences*, **350**, 133–141.

Jablonka, E. & Raz, G. (2009). Transgenerational epigenetic inheritance: prevalence, mechanisms, and implications for the study of heredity and evolution. *Quarterly Review of Biology*, **84**, 131–176.

Jacob, F. & Monod, J. (1961). On the regulation of gene activity. *Cold Spring Harbor Symposium on Quantitative Biology*, **26**, 193–211.

Jahoor, F., Badaloo, A., Reid, M. & Forrester, T. (2006). Unique metabolic characteristics of the major syndromes of severe childhood malnutrition. In *The Tropical Metabolism Research Unit, The University of the West Indies, Jamaica 1956–2006: The House that John Built*, ed. T. Forrester, D. Picou & S. Walker. Kingston, Jamaica: Ian Randle Publishers, 23–60.

Jenuwein, T. & Allis, C. D. (2001). Translating the histone code. *Science*, **293**, 1074–1080.

Johannsen, W. (1909). *Elemente der Exakten Erblichkeitslehre*, Jena, Germany: Gustav Fischer.

Johnson, L. J. & Tricker, P. J. (2010). Epigenomic plasticity within populations: its evolutionary significance and potential. *Heredity*, **105**, 113–121.

Jolly, A. (1966). Lemur social behavior and primate intelligence. *Science*, **153**, 501–506.

Jones, A. P. & Friedman, M. I. (1982). Obesity and adipocyte abnormalities in offspring of rats undernourished during pregnancy. *Science*, **215**, 1518–1519.

Jones, J. H. (2009). The force of selection on the human life cycle. *Evolution and Human Behavior*, **30**, 305–314.

Jones, P. A. (1999). The DNA methylation paradox. *Trends in Genetics*, **15**, 34–37.

Jones, P. A., Rideout, W. M., Shen, J. C., Spruck, C. H. & Tsai, Y. C. (1992). Methylation, mutation and cancer. *Bioessays*, **14**, 33–36.

Kalbe, M., Eizaguirre, C., Dankert, I., et al. (2009). Lifetime reproductive success is maximized with optimal major histocompatibility complex diversity. *Proceedings of the Royal Society B – Biological Sciences*, **276**, 925–934.

Kandel, E. R. & Schwartz, J. H. (1982). Molecular biology of learning: modulation of transmitter release. *Science*, **218**, 433–443.

Kartavtseva, I. V. & Pavlenko, M. V. (2000). Chromosome variation in the striped field mouse *Apodemus agrarius* (Rodentia, Muridae). *Russian Journal of Genetics*, **36**, 162–174.

Keller, E.F. (2000). *Century of the Gene*, Cambridge, MA: Harvard University Press.

Keller, E.F. (2010). *The Mirage of a Space between Nature and Nurture*, Durham, NC: Duke University Press.

Kidwell, M. G. & Lisch, D. R. (2001). Perspective: Transposable elements, parasitic DNA, and genome evolution. *Evolution*, **55**, 1–24.

King, M. (1993). *Species Evolution: The Role of Chromosome Change*, Cambridge, UK: Cambridge University Press.

Kirschner, M. W. & Gerhart, J. C. (2005). *The plausibility of life: resolving Darwin's dilemma*, New Haven: Yale University Press.

Klopfer, P. & Klopfer, M. (1977). Compensatory responses of goat mothers to their impaired young. *Animal Behaviour*, **25**, 286–291.

Köhler, W. (1925). *The Mentality of Apes*, London and New York: K. Paul, Trench, Trubner & Co.

Koyama, F. C., Chakrabarti, D. & Garcia, C. R. S. (2009). Molecular machinery of signal transduction and cell cycle regulation in *Plasmodium*. *Molecular and Biochemical Parasitology*, **165**, 1–7.

Koziol, M. J. & Rinn, J. L. (2010). RNA traffic control of chromatin complexes. *Current Opinion in Genetics & Development*, **20**, 142–148.

Kruuk, L. E. B., Clutton-Brock, T. H., Rose, K. E. & Guinness, F. E. (1999). Early determinants of lifetime reproductive success differ between the sexes in red deer. *Proceedings of the Royal Society of London, Series B, Biological Sciences*, **266**, 1655–1661.

Kucharski, R., Maleszka, J., Foret, S. & Maleszka, R. (2008). Nutritional control of reproductive status in honeybees via DNA methylation. *Science*, **319**, 1827–1830.

Kurth, R. & Bannert, N. (eds) (2010). *Retroviruses: Molecular Biology, Genomics and Pathogenesis*, Norfolk, UK: Caister Academic Press.

Kuzawa, C. W. (2005). Fetal origins of developmental plasticity: are fetal cues reliable predictors of future nutritional environments? *American Journal of Human Biology*, **17**, 5–21.

Lachmann, M. & Jablonka, E. (1996). The inheritance of phenotypes: an adaptation to fluctuating environments. *Journal of Theoretical Biology*, **181**, 1–9.

Laforsch, C., Beccara, L. & Tollrian, R. (2006). Inducible defenses: the relevance of chemical alarm cues in *Daphnia*. *Limnology and Oceanography*, **51**, 1466–1472.

Laland, K. N. & Galef, B. G. (eds) (2009). *The Question of Animal Culture*, Cambridge, MA: Harvard University Press.

Laland, K. N., Odling-Smee, J. & Myles, S. (2010). How culture shaped the human genome: bringing genetics and the human sciences together. *Nature Reviews Genetics*, **11**, 137–148.

Lee, T. M. & Zucker, I. (1988). Vole infant development is influenced perinatally by maternal photoperiodic history. *American Journal of Physiology*, **255**, R831-R38.

Lehrman, D. S. (1970). Semantic and conceptual issues in the nature–nurture problem. In *Development and Evolution of Behavior*, ed. L. R. Aronson, E. Tobach, D. S. Lehrman & J. S. Rosenblatt. San Francisco: Freeman, 17–52.

Lelievre-Pegorier, M., Vilar, J., Ferrier, M.-L., et al. (1998). Mild vitamin A deficiency leads to inborn nephron deficit in the rat. *Kidney International*, **54**, 1455–1462.

Lerner, R. (1984). *On The Nature of Human Plasticity*, Cambridge, UK: Cambridge University Press.

Leroi, A. M. (2003). *Mutants: On the Form, Varieties and Errors of the Human Body*, London: HarperCollins.

Levine, S. (1957). Infantile experience and resistance to physiological stress. *Science*, **126**, 405.

Levine, S. (1969). An endocrine theory of infantile stimulation. In *Stimulation in Early Infancy*, ed. A. Ambrose. London: Academic Press, 45–63.

Lewontin, R. C. (1978). Adaptation. *Scientific American*, **239**, 212–231.

Lewontin, R. C. (1983). Gene, organism and environment. In *Evolution from Molecules to Men*, ed. D. S. Bendall. Cambridge, UK: Cambridge University Press, 273–285.

Li, E. & Bird, A. (2007). DNA methylation in mammals. In *Epigenetics*, ed. C. D. Allis, T. Jenuwein & D. Reinberg. New York: Cold Spring Harbor Laboratory Press, 341–356.

Li, X., Cassidy, J. J., Reinke, C. A., Fischboeck, S. & Carthew, R. W. (2009). A microRNA imparts robustness against environmental fluctuation during development. *Cell*, **137**, 273–282.

Lillycrop, K. A., Slater-Jefferies, J. L., Hanson, M. A., Godfrey, K. M., Jackson, A. A. & Burdge, G. C. (2007). Induction of altered epigenetic regulation of the hepatic glucocorticoid receptor in the offspring of rats fed a protein-restricted diet during pregnancy suggests that reduced DNA methyltransferase-1 expression is involved in impaired DNA methylation and changes in histone modifications. *British Journal of Nutrition*, **97**, 1064–1073.

Linquist, S., Machery, E., Griffiths, P. E. & Stotz, K. (2011). Exploring the folkbiological conception of human nature. *Philosophical Transactions of the Royal Society of London, Series B, Biological Sciences*, **366**, 444–453.

Liu, D., Diorio, J., Tannenbaum, B., et al. (1997). Maternal care, hippocampal glucocorticoid receptors, and hypothalamic-pituitary-adrenal responses to stress. *Science*, **277**, 1659–1662.

Liu, S., Yeh, C.-T., Ji, T., et al. (2009). *Mu* transposon insertion sites and meiotic recombination events co-localize with epigenetic marks for open chromatin across the maize genome. *PLoS Genetics*, **5**, e1000733.

Liu, Y. L., Jia, W. G., Gu, Q, & Cynader, M. (1994). Involvement of muscarinic actetyl-choline receptors in regulation of kitten visual-cortex plasticity. *Brain Research. Developmental Brain Research*, **79**, 63–71.

Lloyd, G. E. R. (2007). *Cognitive Variations: Reflections on the Unity and Diversity of the Human Mind*, Oxford: Oxford University Press.

Lloyd Morgan, C. (1896). *Habit and Instinct*, London: Arnold.

Lorenz, K. (1935). Der kumpan in der umwelt des vogels. *Journal für Ornithologie*, **83**, 137–213, 289–413.

Lorenz, K. (1965). *Evolution and Modification of Behavior*, Chicago: University of Chicago Press.

Lorincz, M. C., Dickerson, D. R., Schmitt, M. & Groudine, M. (2004). Intragenic DNA methylation alters chromatin structure and elongation efficiency in mammalian cells. *Nature Structural & Molecular Biology*, **11**, 1068–1075.

Lucas, A. & Sampson, H. A. (2006). Infant nutrition and primary prevention: Current and future perspectives. In *Primary Prevention by Nutrition Intervention in Infancy and Childhood*, ed. A. Lucas & H. A. Sampson. Basel: Karger, 1–13.

Maher, B. (2008). Personal genomes: the case of the missing heritability. *Nature*, **456**, 18–21.

Male, D., Brostoff, J., Roth, D. B. & Roitt, I. (2006). *Immunology*, Edinburgh: Elsevier.

Maleszka, R. (2008). Epigenetic integration of environmental and genomic signals in honey bees: the critical interplay of nutritional, brain and reproductive networks. *Epigenetics*, **3**, 188–192.

Mameli, M. (2005). The inheritance of features. *Biology & Philosophy*, **20**, 365–399.

Mameli, M. & Bateson, P. (2006). Innateness and the sciences. *Biology & Philosophy*, **21**, 155–188.

Mameli, M. & Bateson, P. (2011). An evaluation of the concept of innateness. *Philosophical Transactions of the Royal Society of London, Series B, Biological Sciences*, **366**, 436–443.

Mancini, D., Singh, S., Ainsworth, P. & Rodenhiser, D. (1997). Constitutively methylated CpG dinucleotides as mutation hot spots in the retinoblastoma gene (RB1). *American Journal of Human Genetics*, **61**, 80–87.

Marler, P. (2004). Innateness and the instinct to learn. *Anais da Academia Brasileira de Ciências*, **76**, 189–200.

Marler, P. & Peters, S. (1977). Selective vocal learning in a sparrow. *Science*, **198**, 519–521.

Marler, P. & Slabbekoorn, H. (2004). *Nature's Music: The Science of Birdsong*, London: Elsevier Academic.

Marshall, D. J. & Uller, T. (2007). When is a maternal effect adaptive? *Oikos*, **116**, 1957–1963.

Martin, P., & Bateson, P. (2007). *Measuring Behaviour*, 3rd edition. Cambridge, UK: Cambridge University Press.

Martin, P. & Caro, T. M. (1985). On the functions of play and its role in behavioral development. *Advances in the Study of Behavior*, **15**, 59–103.

Mateo, J. M. (2007). Ecological and hormonal correlates of antipredator behavior in adult Belding's ground squirrels (*Spermophilus beldingi*). *Behavioral Ecology and Sociobiology*, **62**, 37–49.

Mathis, A., Ferrari, M. C. O., Windel, N., Messier, F. O. & Chivers, D. P. (2008). Learning by embryos and the ghost of predation future. *Proceedings of the Royal Society of London, Series B, Biological Sciences*, **275**, 2603–2607.

Mattick, J. S. (2010). RNA as the substrate for epigenome–environment interactions. *Bioessays*, **32**, 548–552.

Matzke, M. A., Mette, M. F. & Matzke, A. J. M. (2000). Transgene silencing by the host genome defense: implications for the evolution of epigenetic control mechanisms in plants and vertebrates. *Plant Molecular Biology*, **43**, 401–415.

Maynard Smith, J. (1982). *Evolution and the Theory of Games*, Cambridge, UK: Cambridge University Press.

Maynard Smith, J. & Szathmáry, E. (1995). *The Major Transitions in Evolution*, New York: Oxford University Press.

Mayo, A. E., Setty, Y., Shavit, S., Zaslaver, A. & Alon, U. (2006). Plasticity of the *cis*-regulatory input function of a gene. *PLoS Biology*, **4**, 555–561.

Mayr, E. (1963). *Animal Species and Evolution*, Cambridge, MA: Harvard University Press.

McGrath, B. (2007). Muscle memory: the next generation of bionic prostheses. *New Yorker*, 30 July 2007, 40–45.

Meaney, M. J. (2001). Maternal care, gene expression, and the transmission of individual differences in stress reactivity across generations. *Annual Review of Neuroscience*, **24**, 1161–1192.

Meaney, M. J. (2010). Epigenetics and the biological definition of gene × environment interactions. *Child Development*, **81**, 41–79.

Miles, J. L., Landon, J., Davison, M., Krageloh, C. U., Thompson, N. M., Triggs, C. M. & Breier, B. H. (2009). Prenatally undernourished rats show increased preference for wheel running v. lever pressing for food in a choice task. *British Journal of Nutrition*, **101**, 902–908.

Milinski, M. (1999). Glasses for children: are they curing the wrong symptoms for the wrong reason? In *Evolution in Health and Disease*, ed. S. C. Stearns. Oxford: Oxford University Press, 121.

Miller, C. A., Gavin, C. F., White, J. A., et al. (2010). Cortical DNA methylation maintains remote memory. *Nature Neuroscience*, **13**, 664–666.

Minelli, A. & Fusco, G. (2010). Developmental plasticity and the evolution of animal complex life cycles. *Philosophical Transactions of the Royal Society of London, Series B, Biological Sciences*, **365**, 631–640.

Misawa, K., Kamatani, N. & Kikuno, R. F. (2008). The universal trend of amino acid gain-loss is caused by CpG hypermutability. *Journal of Molecular Evolution*, **67**, 334–342.

Misawa, K. & Kikuno, R. F. (2009). Evaluation of the effect of CpG hypermutability on human codon substitution. *Gene*, **431**, 18–22.

Mohd-Sarip, A. & Verrijzer, C. P. (2004). A higher order of silence. *Science*, **306**, 1484–1485.

Monaghan, P. (2008). Early growth conditions, phenotypic development and environmental change. *Philosophical Transactions of the Royal Society of London, Series B, Biological Sciences*, **363**, 1635–1645.

Mondal, T., Rasmussen, M., Pandey, G. K., Isaksson, A. & Kanduri, C. (2010). Characterization of the RNA content of chromatin. *Genome Research*, **20**, 899–907.

Moore, B R. (2004). The evolution of learning. *Biological Reviews*, **79**, 301–335.

Moran, N. A. (1992). The evolutionary maintenance of alternative phenotypes. *American Naturalist*, **139**, 971–989.

Moss, L. (2002). *What Genes Can't Do*, Cambridge, MA: MIT Press.

Mrosovsky, N. (1990). *Rheostasis: The Physiology of Change*, New York: Oxford University Press.

Murgatroyd, C., Patchev, A. V., Wu, Y., et al. (2009). Dynamic DNA methylation programs persistent adverse effects of early-life stress. *Nature Neuroscience*, **12**, 1559–1566.

Nadeau, J. H. (2009). Transgenerational genetic effects on phenotypic variation and disease risk. *Human Molecular Genetics*, **18**, R202–R210.

Nelson, V. R., Spiezio, S. H. & Nadeau, J. H. (2010). Transgenerational genetic effects of the paternal Y chromosome on daughters' phenotypes. *Epigenomics*, **2**, 513–521.

Nettle, D., Coall, D. A. & Dickins, T. E. (2010). Birthweight and paternal involvement predict early reproduction in British women: evidence from the National Child Development Study. *American Journal of Human Biology*, **22**, 172–179.

Neuberger, M. S. (2008). Antibody diversification by somatic mutation: from Burnet onwards. *Immunology & Cell Biology*, **86**, 124–132.

Newman, S. A. (2007). William Bateson's physicalist ideas. In *From Embryology to Evo-Devo: A History of Evolutionary Development*, ed. M. Laubichler & J. Maienschein. Cambridge, MA: MIT Press,

Nijhout, H. F. (2002). The nature of robustness in development. *Bioessays*, **24**, 553–563.

Nishimura, K. (2006). Inducible plasticity: optimal waiting time for the development of an inducible phenotype. *Evolutionary Ecology Research*, **8**, 553–559.

Oberlander, T. F., Weinberg, J., Papsdorf, M., Grunau, R., Misri, S. & Devlin, A. M. (2008). Prenatal exposure to maternal depression, neonatal methylation of human glucocorticoid receptor gene (*NR3C1*) and infant cortisol stress responses. *Epigenetics*, **3**, 97–106.

Odling-Smee, F. J., Laland, K. N. & Feldman, M. W. (2003). *Niche Construction: The Neglected Process of Evolution*, Princeton, NJ: Princeton University Press.

Osborn, H. F. (1896). A mode of evolution requiring neither natural selection nor the inheritance of acquired characters. *Transactions of the New York Academy of Sciences*, **15**, 141–142, 148.

Oyama, S. (2000). *The Ontogeny of Information: Developmental Systems and Evolution*, 2nd edition, Durham, NC: Duke University Press.

Oyama, S., Griffiths, P. E. & Gray, R. D. (2001). *Cycles of Contingency: Developmental Systems and Evolution*, Cambridge, MA: MIT Press.

Paenke, I., Kawecki, T. J. & Sendhoff, B. (2009). The influence of learning on evolution: a mathematical framework. *Artificial Life*, **15**, 227–245.

Painter, R. C., Roseboom, T. J. & Bleker, O. P. (2005). Prenatal exposure to the Dutch famine and disease in later life: an overview. *Reproductive Toxicology*, **20**, 345–352.

Paley, W. (1802). *Natural Theology: Or, Evidences of The Existence and Attributes of The Deity, Collected From The Appearances of Nature*, London: Faulder.

Paz-Yaacov, N., Levanon, E. Y., Nevo, E., et al. (2010). Adenosine-to-inosine RNA editing shapes transcriptome diversity in primates. *Proceedings of the National Academy of Sciences of the USA*, **107**, 12174–12179.

Peleg, S., Sananbenesi, F., Zovoilis, A., et al. (2010). Altered histone acetylation is associated with age-dependent memory impairment in mice. *Science*, **328**, 753–756.

Pellegrini, A. D. (2009). *The Role of Play in Human Development*, Oxford: Oxford University Press.

Pepperberg, I. M. (2008). *Alex and Me: How a Scientist and a Parrot Discovered a Hidden World of Animal Intelligence – and Formed a Deep Bond in The Process*, New York: Collins.

Perera, F., Tang, W.-Y., Herbstman, J., et al. (2009). Relation of DNA methylation of 5'-CpG island of *ACSL3* to transplacental exposure to airborne polycyclic aromatic hydrocarbons and childhood asthma. *PLoS One*, **4**, e4488.

Perry, M. W., Boettiger, A. N., Bothma, J. P., & Levine, M. (2010). Shadow enhancers foster robustness of *Drosophila* gastrulation. *Current Biology*, **20**, 1562–1567.

Pfeifer, G. P. (2006). Mutagenesis at methylated CpG sequences. *Current Topics in Microbiology and Immunology*, **301**, 259–281.

Pfennig, D. W. & McGee, M. (2010). Resource polyphenism increases species richness: a test of the hypothesis. *Philosophical Transactions of the Royal Society of London. Series B, Biological Sciences*, **365**, 577–591.

Pfennig, D. W., Wund, M. A., Snell-Rood, E. C., Cruickshank, T., Schlichting, C. D. & Moczek, A. P. (2010). Phenotypic plasticity's impacts on diversification and speciation. *Trends in Ecology and Evolution*, **25**, 459–467.

Piaget, J. (1979). *Behaviour and Evolution*, London: Routledge & Kegan Paul.

Pigliucci, M. (2001). *Phenotypic Plasticity: Beyond Nature and Nurture*, Baltimore: Johns Hopkins University Press.

Pigliucci, M. (2010). Genotype–phenotype mapping and the end of the 'genes as blueprint' metaphor. *Philosophical Transactions of the Royal Society of London. Series B, Biological Sciences*, **365**, 557–566.

Pigliucci, M. & Müller, G. B. (eds) (2010). *Evolution: The Extended Synthesis*, Cambridge, MA: MIT Press.

Pigliucci, M. & Murren, C. J. (2003). Genetic assimilation and a possible evolutionary paradox: can macroevolution sometimes be so fast as to pass us by? *Evolution*, **57**, 1455–1464.

Plagemann, A., Harder, T., Brunn, M., et al. (2009). Hypothalamic proopiomelanocortin promoter methylation becomes altered by early overfeeding: an epigenetic model of obesity and the metabolic syndrome. *Journal of Physiology*, **587**, 4963–4976.

Polosina, Y. Y. & Cupples, C. G. (2010). MutL: conducting the cell's response to mismatched and misaligned DNA. *Bioessays*, **32**, 51–59.

Price, T. D., Qvarnström, A. & Irwin, D. E. (2003). The role of phenotypic plasticity in driving genetic evolution. *Proceedings of the Royal Society of London, Series B, Biological Sciences*, **270**, 1433–1440.

Rampon, C., Tang, Y.-P., Goodhouse, J., Shimizu, E., Kyin, M. & Tsien, J. Z. (2000). Enrichment induces structural changes and recovery from nonspatial memory deficits in CA1 NMDAR1-knockout mice. *Nature Neuroscience*, 3, 238–244.

Rassoulzadegan, M., Grandjean, V., Gounon, P., Vincent, S., Gillot, I. & Cuzin, F. (2006). RNA-mediated non-Mendelian inheritance of an epigenetic change in the mouse. *Nature*, **441**, 469–474.

Rauschecker, J. P. & Marler, P. (eds) (1987). *Imprinting and Cortical Plasticity: Comparative Aspects of Sensitive Periods*, New York: John Wiley & Sons.

Reader, S. M. & Laland, K. N. (2002). Social intelligence, innovation, and enhanced brain size in primates. *Proceedings of the National Academy of Sciences of the USA*, 99, 4436–41.

Remy, J. J. & Hobert, O. (2005). An interneuronal chemoreceptor required for olfactory imprinting in *C. elegans*. *Science*, **309**, 787–790.

Ridley, M. (2003). *Nature via Nurture: Genes, Experience, and What Makes Us Human*, New York: HarperCollins.

Roberts, T. F., Tschida, K. A., Klein, M. E. & Mooney, R. (2010). Rapid spine stabilization and synaptic enhancement at the onset of behavioural learning. *Nature*, **463**, 948–952.

Robinson, B.W. & Dukas, R. (1999). The influence of phenotypic modification on evolution: the Baldwin effect and modern perspectives. *Oikos*, **85**, 582–589.

Rosen, J. M. & Jordan, C. T. (2009). The increasing complexity of the cancer stem cell paradigm. *Science*, **324**, 1670–1673.

Rosenfield, R. L., Cooke, D. W. & Radovick, S. (2008). Puberty and its disorders in the female. In *Pediatric Endocrinology*, 3rd edition, ed. M. Sperling. Philadelphia, PA: Saunders, 530–609.

Rowell, C. H. F. (1971). The variable coloration of the Acridoid grasshoppers. *Advances in Insect Physiology*, **8**, 145–198.

Rutherford, S. L. & Lindquist, S. (1998). Hsp90 as a capacitor for morphological evolution. *Nature*, **396**, 336–342.

Sameroff, A. (2010). A unified theory of development: a dialectic integration of nature and nurture. *Child Development*, **81**, 6–22.

Scarr, S. & McCartney, K. (1983). How people make their own environments: a theory of genotype → environment effects. *Child Development*, **54**, 424–435.

Schlichting, C. D. & Pigliucci, M. (1998). *Phenotypic Evolution: A Reaction Norm Perspective*, Sunderland, MA: Sinauer Associates.

Schmalhausen, I. I. (1949). *Factors of Evolution*, Philadelphia, PA: Blakiston.

Schorderet, D. F. & Gartler, S. M. (1992). Analysis of CpG suppression in methylated and nonmethylated species. *Proceedings of the National Academy of Sciences of the USA*, **89**, 957–961.

Schwarz, R. H. & Yaffe, S. J. (Eds.) (1980). *Drug and Chemical Risks to the Fetus and Newborn*, New York: Alan Liss.

Segerstrom, S. C. (2007). Stress, energy, and immunity. *Current Directions in Psychological Science*, **16**, 326–330.

Sgrò, C. M., Wegener, B. & Hoffmann, A. A. (2010). A naturally occurring variant of Hsp90 that is associated with decanalization. *Proceedings of the Royal Society of London, Series B, Biological Sciences*, **277**, 2049–2057.

Shenk, D. (2010). *The Genius in All of Us*, New York: Doubleday.

Sheriff, M. J., Krebs, C. J. & Boonstra, R. (2009). The sensitive hare: sublethal effects of predator stress on reproduction in snowshoe hares. *Journal of Animal Ecology*, **78**, 1249–1258.

Sheriff, M. J., Krebs, C. J. & Boonstra, R. (2010). The ghosts of predators past: population cycles and the role of maternal programming under fluctuating predation risk. *Ecology*, **91**, 2982–2994.

Sherratt, T. N. (2002). The coevolution of warning signals. *Proceedings of the Royal Society of London, Series B, Biological Sciences*, **269**, 741–746.

Shettleworth, S. J. (2010). *Cognition, Evolution and Behavior*, 2nd edition, New York: Oxford University Press.

Simpson, G. G. (1953). *The Major Features of Evolution*, New York: Columbia University Press.

Slatkin, M. (1974). Hedging one's evolutionary bets. *Nature*, **250**, 704–705.

Slijper, E. J. (1942). Biologic-anatomical investigations on the bipedal gait and upright posture in mammals, with special reference to a little goat, born without forelegs. I and II. *Proceedings of the Koninklijke Nederlandse Akademie Wetenschappen*, **45**, 288–295, 407–415.

Sloboda, D. M., Hart, R., Doherty, D. A., Pennell, C. E. & Hickey, M. (2007). Age at menarche: influences of prenatal and postnatal growth. *Journal of Clinical Endocrinology and Metabolism*, **92**, 46–50.

Sloboda, D. M., Howie, G. J., Pleasants, A., Gluckman, P. D. & Vickers, M. H. (2009). Pre- and postnatal nutritional histories influence reproductive maturation and ovarian function in the rat. *PLoS One*, **4**, e6744.

Slotkin, R. K. & Martienssen, R. (2007). Transposable elements and the epigenetic regulation of the genome. *Nature Reviews Genetics*, **8**, 272–285.

Smyth, M. J., Dunn, G. P. & Schreiber, R. D. (2006). Cancer immunosurveillance and immunoediting: the roles of immunity in suppressing tumor development and shaping tumor immunogenicity. *Advances in Immunology*, **90**, 1–50.

Sokolov, E. N. (1963). *Perception and the Conditioned Reflex*, Oxford: Pergamon.

Sol, D., Duncan, R. P., Blackburn, T. M., Cassey, P. & Lefebvre, L. (2005). Big brains, enhanced cognition, and response of birds to novel environments. *Proceedings of the National Academy of Sciences*, **102**, 5460–5465.

Sollars, V., Lu, X., Xiao, L., Wang, X., Garfinkel, M. D. & Ruden, D. M. (2003). Evidence for an epigenetic mechanism by which Hsp90 acts as a capacitor for morphological evolution. *Nature Genetics*, **33**, 70–74.

Spalding, D. A. (1873). Instinct: with original observations on young animals. *Macmillan's Magazine*, **27**, 282–293 (Reprinted in 1954 in the *British Journal of Animal Behaviour*, **2**, 1–11).

Spemann, H. (1938). *Embryonic Development and Induction*, New Haven, NJ: Yale University Press.

Spencer, H. G., Hanson, M. A. & Gluckman, P. D. (2006). Response to Wells: phenotypic responses to early environmental cues can be adaptive in adults. *Trends in Ecology & Evolution*, **21**, 425–426.

Standing, L. (1973). Learning 10,000 pictures. *Quarterly Journal of Experimental Psychology*, **25**, 207–22.

Stockard, C. R. (1921). Developmental rate and structural expression: an experimental study of twins, 'double monsters' and single deformities, and the interaction among embryonic organs during their origin and development. *American Journal of Anatomy*, **28**, 115–275.

Stocum, D. L. (2006). *Regenerative Biology and Medicine*, New York: Elsevier.

Stouder, C. & Paoloni-Giacobino, A. (2010). Transgenerational effects of the endocrine disruptor vinclozolin on the methylation pattern of imprinted genes in the mouse sperm. *Reproduction*, **139**, 373–379.

Suemori, H. & Noguchi, S. (2000). Hox C cluster genes are dispensable for overall body plan of mouse embryonic development. *Developmental Biology*, **220**, 333–342.

Sultan, S. E., Barton, K. & Wilczek, A. M. (2009). Contrasting patterns of transgenerational plasticity in ecologically distinct congeners. *Ecology*, **90**, 1831–1839.

Sultan, S. E. & Spencer, H. G. (2002). Metapopulation structure favors plasticity over local adaptation. *American Naturalist*, **160**, 271–283.

Suzuki, R. & Arita, T. (2007). The dynamic changes in roles of learning through the Baldwin effect. *Artificial Life*, **13**, 31–43.

Suzuki, Y. & Nijhout, H. F. (2006). Evolution of a polyphenism by genetic accommodation. *Science*, **311**, 650–652.

Suzuki, Y. & Nijhout, H. F. (2008). Genetic basis of adaptive evolution of a polyphenism by genetic accommodation. *Journal of Evolutionary Biology*, **21**, 57–66.

Tahiliani, M., Koh, K. P., Shen, Y. H., et al. (2009). Conversion of 5-methylcytosine to 5-hydroxymethylcytosine in mammalian DNA by MLL partner TET1. *Science*, **324**, 930–935.

Tautz, J. (2008). *The Buzz about Bees: Biology of a Superorganism*, Berlin: Springer.

Tebbich, S., Sterelny, K. & Teschke, I. (2010). The tale of the finch: adaptive radiation and behavioural flexibility. *Philosophical Transactions of the Royal Society of London, Series B, Biological Sciences*, **365**, 1099–1109.

Tebbich, S., Taborsky, M., Fessl, B. & Blomquist, D. (2001). Do woodpecker finches acquire tool-use by social learning? *Proceedings of the Royal Society of London, Series B, Biological Sciences*, 268, 2189–2193.

ten Cate, C. (1986). Does behavior contingent stimulus movement enhance filial imprinting in Japanese quail? *Developmental Psychobiology*, 19, 607–614.

The Chimpanzee Sequencing and Analysis Consortium (2005). Initial sequence of the chimpanzee genome and comparison with the human genome. *Nature*, 437, 69–87.

Thelen, E. (1989). Self-organization in developmental processes: Can systems approaches work? In *Systems and Development: The Minnesota Symposia on Child Psychology. Vol. 22*, ed. M. R. Gunnar & E. Thelen. Hillsdale, NJ: Erlbaum, 77–117.

Thoman, E. B. & Levine, S. (1970). Hormonal and behavioral changes in the rat mother as a function of early experience treatments of the offspring. *Physiology & Behavior*, 5, 1417–21.

Thompson, J. D. (1991). Phenotypic plasticity as a component of evolutionary change. *Trends in Ecology & Evolution*, 6, 246–249.

Thorndike, E. L. (1898). Animal intelligence. *Psychological Review*, Supplement 2, 1–109.

Thorpe, W. H. (1956). *Learning and Instinct in Animals*, London: Methuen.

Tieri, P., Castellani, G., Remondini, D., Valensin, S., Loroni, J., Salvioli, S. & Franceschi, C. (2007). Capturing degeneracy in the immune system. In *In Silico Immunology*, ed. D. D. R. Flower and J. Timmis. New York: Springer, 109–118.

Tinbergen, N. (1963). On aims and methods of ethnology. *Zeitschrift für Tierpsychologie*, 20, 410–433.

Trivers, R. L. (1974). Parent–offspring conflict. *American Zoologist*, 14, 249–264.

True, H. L. & Lindquist, S. L. (2000). A yeast prion provides a mechanism for genetic variation and phenotypic diversity. *Nature*, 407, 477–483.

Trut, L., Oskina, I. & Kharlamova, A. (2009). Animal evolution during domestication: the domesticated fox as a model. *Bioessays*, 31, 349–360.

Vajo, Z., Francomano, C. A. & Wilkin, D. J. (2000). The molecular and genetic basis of fibroblast growth factor receptor 3 disorders: the achondroplasia family of skeletal dysplasias, Muenke craniosynostosis, and Crouzon syndrome with acanthosis nigricans. *Endocrine Reviews*, 21, 23–39.

Vallortigara, G., Regolin, L. & Marconato, F. (2005). Visually inexperienced chicks exhibit spontaneous preference for biological motion patterns. *PLoS Biology*, 3, 1–5.

Venditti, C., Meade, A. & Pagel, M. (2010). Phylogenies reveal new interpretation of speciation and the Red Queen. *Nature*, 463, 349–352.

Vickers, M. H., Breier, B. H., Cutfield, W. S., Hofman, P. L. & Gluckman, P. D. (2000). Fetal origins of hyperphagia, obesity, and hypertension and postnatal amplification by hypercaloric nutrition. *American Journal of Physiology*, 279, E83–E87.

Vickers, M. H., Breier, B. H., McCarthy, D. & Gluckman, P. D. (2003). Sedentary behavior during postnatal life is determined by the prenatal environment and exacerbated by postnatal hypercaloric nutrition. *American Journal of Physiology*, 285, R271–R273.

Vickers, M. H., Gluckman, P. D., Coveny, A. H., et al. (2005). Neonatal leptin treatment reverses developmental programming. *Endocrinology*, 146, 4211–4216.

von Bertalanffy, L. (1974). *Perspectives on General System Theory*, New York: Brazillier.

von Moltke, H. J. & Olbing, H. (1989). Die Ausbildungs – und Berufssituation contergangeschadigter junger Erwachsener. *Rehabilitation*, 28, 78–82.

Waddington, C. H. (1953). Genetic assimilation of an acquired character. *Evolution*, 7, 118–126.

Waddington, C. H. (1957). *The Strategy of the Genes*, London: Allen & Unwin.

Waddington, C. H. (1959). Evolutionary systems: animal and human. *Nature*, **183**, 1634–1638.

Wagner, A. (2008). Gene duplications, robustness and evolutionary innovations. *Bioessays*, **30**, 367–373.

Wagner, A. R. (1981). SOP: a model of automatic memory processing in animal behavior. In *Information Processing in Animal Behavior*, ed. W. E. Spear & R. R. Miller. Hillsdale, NJ: Erlbaum, 5–47.

Wagner, G. P. & Altenberg, L. (1996). Complex adaptations and the evolution of evolvability. *Evolution*, **50**, 967–976.

Wagner, G. P., Pavlicev, M. & Cheverud, J. M. (2007). The road to modularity. *Nature Reviews Genetics*, **8**, 921–931.

Wagner, K. D., Wagner, N., Ghanbarian, H., et al. (2008). RNA induction and inheritance of epigenetic cardiac hypertrophy in the mouse. *Developmental Cell*, **14**, 962–969.

Wake, G. C., Pleasants, A. B., Beedle, A. S. & Gluckman, P. D. (2010). A model for phenotype change in a stochastic framework. *Mathematical Biosciences and Engineering*, **7**, 721–730.

Walker, R., Gurven, M., Hill, K., et al. (2006). Growth rates and life histories in twenty-two small-scale societies. *American Journal of Human Biology*, **18**, 295–311.

Wallace, B. (1986). Can embryologists contribute to an understanding of evolutionary mechanisms? In *Integrating Scientific Disciplines*, ed. W. Bechtel. Dordrecht, The Netherlands: Martinus Nijhoff,

Wallace, M. R., Andersen, L. B., Saulino, A. M., Gregory, P. E., Glover, T. W. & Collins, F. S. (1991). A *de novo Alu* insertion results in neurofibromatosis type 1. *Nature*, **353**, 864–866.

Walser, J.-C. & Furano, A. V. (2010). The mutational spectrum of non-CpG DNA varies with CpG content. *Genome Research*.

Walser, J.-C., Ponger, L. C. & Furano, A. V. (2008). CpG dinucleotides and the mutation rate of non-CpG DNA. *Genome Research*, **18**, 1403–1414.

Walton, A. J. & Hammond, J. (1938). The maternal effects on growth and conformation in Shire horses–Shetland pony crosses. *Proceedings of the Royal Society of London, Series B, Biological Sciences*, **125**, 311–335.

Wang, H.-Y., Tang, H., Shen, C. K. J. & Wu, C.-I. (2003). Rapidly evolving genes in human. I. The glycophorins and their possible role in evading malaria parasites. *Molecular Biology & Evolution*, **20**, 1795–1804.

Waterland, R. A. & Jirtle, R. L. (2003). Transposable elements: targets for early nutritional effects on epigenetic gene regulation. *Molecular and Cellular Biology*, **23**, 5293–5300.

Weaver, I. C. G., Cervoni, N., Champagne, F. A., et al. (2004). Epigenetic programming by maternal behavior. *Nature Neuroscience*, **7**, 847–854.

Weaver, I. C. G., Champagne, F. A., Brown, S. E., et al. (2005). Reversal of maternal programming of stress responses in adult offspring through methyl supplementation: altering epigenetic marking later in life. *Journal of Neuroscience*, **25**, 11045–11054.

Weaver, I. C. G., D'Alessio, A. C., Brown, S. E., et al. (2007). The transcription factor nerve growth factor-inducible protein A mediates epigenetic programming: altering epigenetic marks by immediate-early genes. *Journal of Neuroscience*, **27**, 1756–1768.

Weismann, A. (1885). *Die Kontinuität des Keimplasmas als Grundlage einer Theorie der Vererbung*, Jena, Germany: Gustav Fischer.

Wells, J. C. K. (2006). Is early development in humans a predictive adaptive response anticipating the adult environment? *Trends in Ecology & Evolution*, **21**, 424–425.

Wells, M. J. (1967). Sensitization and the evolution of associative learning. In *Symposium on Neurobiology of Invertebrates*, ed. J. Salanki. New York: Plenum, 391–411.

West-Eberhard, M. J. (2003). *Developmental Plasticity and Evolution*, New York: Oxford University Press.

West-Eberhard, M. J. (2005a). Developmental plasticity and the origin of species differences. *Proceedings of the National Academy of Sciences of the USA*, **102**, 6543–6549.

West-Eberhard, M. J. (2005b). Phenotypic accommodation: adaptive innovation due to developmental plasticity. *Journal of Experimental Zoology, Part B, Molecular and Developmental Evolution*, **304B**, 610–618.

Whimbey, A. E. & Denenberg, V. H. (1967). Experimental programming of life histories: factor structure underlying experimentally created individual differences. *Behaviour*, **29**, 296–314.

Wilson, E. O. (1971). Social Insects. *Science*, **172**, 406.

Wilson, E. O. (1975). *Sociobiology: The New Synthesis*, Cambridge, MA: Harvard University Press.

Wilson, E. O. (1976). Author's reply to multiple review of Wilson's *Sociobiology*. *Animal Behaviour*, **24**, 716–718.

Wolterek, R. (1909). Weitere experimentelle untersüchungen über Artveränderung, speziell über das wesen quantitativer Artunterschiede bei Daphniden. *Verhandlungen de Deutschen Zooligischen Gesellschaft*, **19**, 110–172.

Wong, C. C. Y., Caspi, A., Williams, B., et al. (2010). A longitudinal study of epigenetic variation in twins. *Epigenetics*, **5**, 1–11.

World Bank (2009). *World Development Report 2009*, Washington, DC: World Bank.

Wright, C. L., Schwarz, J. S., Dean, S. L. & McCarthy, M. M. (2010). Cellular mechanisms of estradiol-mediated sexual differentiation of the brain. *Trends in Endocrinology & Metabolism*, **21**, 553–561.

Wyles, J. S., Kunkel, J. G. & Wilson, A. C. (1983). Birds, behavior, and anatomical evolution. *Proceedings of the National Academy of Sciences of the USA*, **80**, 4394–4397.

Yazbek, S. N., Nadeau, J. H. & Buchner, D. A. (2010). Ancestral paternal genotype controls body weight and food intake for multiple generations. *Human Molecular Genetics*.

Yntema, C. L. & Mrosovsky, N. (1982). Critical periods and pivotal temperatures for sexual differentiation in loggerhead sea turtles. *Canadian Journal of Zoology*, **60**, 1012–1016.

Zuckerkandl, E. & Cavalli, G. (2007). Combinatorial epigenetics, 'junk DNA', and the evolution of complex organisms. *Gene*, **390**, 232–242.

Index

151

Printed in the United States
By Bookmasters